Honored To Serve

"In their own words"

Written by

Wayne Soares

With Mark Lund

For more information, visit www.SummerlandPublishing.com.
Printed in the U. S. A.
Library of Congress #2025903033

Acknowledgements:

To Mark Lund, my friend, thank you for your unwavering dedication, steadfast commitment, and deep passion for ensuring that these powerful stories of our military veterans are told with the honor and respect they deserve.

***

Maria Yarnall, for your excellent cover image.

***

Jolinda Pizzirani, Summerland Publishing
www.summerlandpublishing.com

*** *** ***

A portion of the proceeds from ***Honored to Serve*** will go to various organizations supporting our military veterans.

Dedication

This book is dedicated to my grandfather and hero,

William "Blink" Soares

- United States Navy, Seabees 1943 - 1946.

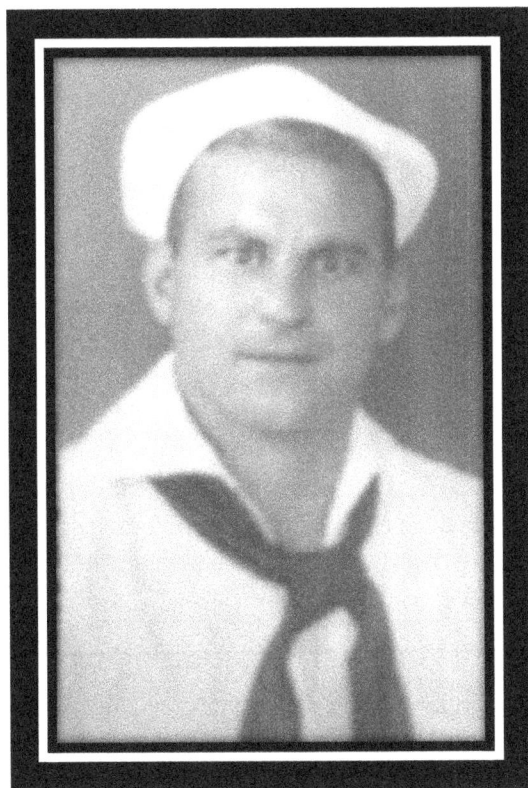

FORWARD

The stories of those who have served in the military are more than just historical accounts—they are personal testaments to courage, sacrifice, and resilience. This collection of interviews, spanning from 1941 to today, offers a rare and powerful window into the experiences of veterans across generations. Through their voices, we gain insight not only into the battles they fought but also into the friendships they forged, the burdens they carried, and the memories—both harrowing and heroic—that have shaped their lives.

At the heart of this remarkable book is Wayne "Wayno" Soares, a man I have had the privilege of calling a close friend for many years. Wayne possesses a rare and invaluable gift: the ability to make people feel comfortable, understood, and truly heard. His engaging personality puts veterans at ease, allowing them to open up in ways they may never have before. Many of the stories shared in these pages—some of the deepest, darkest memories these men and women have carried for decades—have never been spoken aloud until now. That is a testament not only to the bravery of these veterans but also to Wayne's unwavering respect and empathy for those who have served.

History often captures war through facts and figures, strategies and outcomes. But this book does something more—it humanizes military service. It reminds us that behind every uniform is an individual with a unique journey, a family left behind, and a future forever shaped by their experiences. Whether these veterans served in World War II, the Korean War, Vietnam, the Cold War, the Gulf War, or the conflicts of today, their voices form a powerful chorus of sacrifice, resilience, and, in many cases, healing.

For those who have served, these stories will resonate in deeply personal ways. For those who have not, they offer a rare and invaluable glimpse into the realities of military life. And for all of us, they serve as a reminder to listen, to learn, and to honor the men and women who have given so much.

It is an honor to introduce this collection, and I encourage you to read it with an open heart and a deep sense of gratitude. These are not just war stories; they are life stories—stories of bravery, loss, camaraderie, and perseverance. Thanks to Wayne's unique ability to bring them to the surface, we now have the privilege of hearing them.

To my fellow military members for their commitment to the defense of this great nation, and to Wayno for his tireless efforts on behalf of veterans, I proudly say: ***"Thank you for your service!"***

Donald "Q" Quenneville Brigadier General, USAF (Ret.)
March 2025

PROLOGUE

The first time I met a combat veteran and truly listened to their story, I knew my life had changed. I had always admired the men and women in uniform—their discipline, their strength, their unwavering sense of duty. But until I had the privilege of sitting down with them, looking them in the eye, and hearing their stories firsthand, I never truly grasped the depth of what they had endured, not just in battle, but in the years that followed.

Like many Americans, I grew up watching Bob Hope perform for the troops. The way he connected with them, bringing laughter to those who had seen the worst of humanity, left a mark on me. His work planted a seed in my heart and a desire to give back to those who had given so much.

I never served in the military, and that remains one of my biggest regrets. But life has a way of putting us where we are meant to be. In my early 30s, I took a leap into stand-up comedy, hoping that one day, I could use it to bring joy to the men and women who wear the uniform. That dream became a reality when I was working for ESPN Radio and was introduced to Oakland Raiders legend Otis Sistrunk. Our friendship led to an opportunity to perform for the troops at Fort Lewis, Washington, where I was not only able to make them laugh but spent some time with them sharing meals, visiting wounded warriors, and listening to their stories.

It was in those quiet conversations away from the stage, away from the crowds that I discovered something profound: Veterans don't just open up to anyone. Trust is earned. You can't force a conversation about war, about loss, about the struggles of coming home to a country that doesn't always understand what they've been through. They have to feel that you truly care, that you aren't just looking for a story, but that you genuinely *see* them.

One of the first veterans I interviewed was **Private Jerry Anacosta**, who served in World War II. His voice was steady but filled with emotion as he recalled the bitter cold of the Battle of the Bulge, a cold so piercing it was impossible to forget. *"Jesus, it was cold... Christ sakes, you never could imagine THAT type of freezing cold,"* he told me. He went on to describe the harrowing moment when he and his squad had to bury themselves in snow to avoid being spotted by a German convoy. *"If it had been daylight, we'd have been dead for sure."* Even decades later, the memory was fresh in his mind—the fear, the determination, and the incredible resilience it took to survive.

Then there was **Captain Bill Anners**, a helicopter pilot in Vietnam who carried the weight of life-and-death decisions on his shoulders every time he flew a mission. He recounted the moment a sniper's bullet tore through his co-pilot's neck, covering the cockpit in blood. *"He survived, but he lost a lot of blood,"* he said, shaking his head. Another time, after being shot down in the jungle, he and his men waited for hours—hiding, listening, fearing the worst. *"We were scared shitless,"* he admitted, but in the face of danger, his training and camaraderie kept him focused.

Some stories broke my heart. **Sergeant Roger Hampton**, a Bronze Star recipient, reflected on the toll of combat, the relentless fight for survival, and the losses that haunted him. *"All I know is we lost a great many men, and I still sleep like hell thinking about the guys we lost—now more than ever"*. He spoke about spending his first Christmas in a foxhole, gunfire ringing in the distance, unable to think about home because survival was the only thing that mattered.

Through it all, I have been humbled by their courage—not just in war, but in allowing themselves to be vulnerable. To let their stories be told. To trust me with the memories they carry.

Earning that trust became my mission. I didn't just want to write about veterans, I wanted to understand them. Over the years, I've sat with warriors from every branch of service, spanning generations from World War II to Iraq and Afghanistan. Many were hesitant at first. Some had never shared their experiences with anyone outside of their brothers and sisters in arms. Others had buried their memories so deep that bringing them to the surface was painful.

But I learned that trust is built in the small moments. It's in showing up, in listening without judgment, in making sure they know that their voices matter. It's in the shared laughter over a meal, the quiet nod of understanding when a story becomes too difficult to finish. It's in honoring not just their service, but the *person* behind the uniform—the father, the mother, the son, the daughter, the friend.

What I also learned and what too many people don't realize is that for many veterans, *the war never really ends.* PTSD is not just a diagnosis; it's a daily battle. The memories don't fade, the sounds of war don't disappear, and for some, the struggle to reintegrate into civilian life is just as grueling as the battlefield itself. Some have been able to heal, to find new purpose, to move forward, and have become incredibly successful in their pursuits after serving our country.   Others are still fighting demons alone, unheard, sometimes forgotten.

This book is their voice. These are their stories in their own words.

Here, you will meet men and women who have served in every major conflict from World War II to the present day. Their stories are raw, unfiltered, and deeply personal. Some will inspire you. Some will break your heart. But all of them deserve to be told.

I share these profiles not just to honor our veterans, but to remind us of what they have sacrificed for us. Their courage, their resilience, their pain, and their triumphs are part of the fabric of this nation.

So, as you read, I ask one thing of you: **Remember them.**

Remember the 18-year-old who landed on Omaha Beach, terrified but determined.
Remember the pilot who flew through enemy fire
to rescue his brothers.
Remember the soldier who came home but
never truly left the battlefield.
And remember those who never came home at all.

And more than that—**support them.**

If you see a veteran, shake their hand. Thank them. Listen to them. If you have the means, donate to organizations that help veterans with PTSD, homelessness, and job reintegration. If you run a business, hire a veteran. If you know one struggling, be there for them.

Because the words *thank you for your service* mean so much more when followed by action.

I am honored to share these stories. And I will continue to do so for as long as I can.

To the men and women who have served, who continue to serve, and to those who never made it home—**thank you for your service to your country. Your courage, your sacrifice, and your unwavering dedication will never be forgotten.**

Welcome Home.

World War II (1942-1946)

Veterans Spotlights

Private Jerry Anacosta- United States Army, (1942–1945)
Petty Officer Bernie Auge – United States Navy (1942–1946)
Private First Class Sam Baxter- United States Marine Corps (1942–1945)
Private Charlie Barten - United States Marine Corps (1942–1945)
*Seaman Lionel "Red" Blanchard – United States Navy (1941–1952)
Private Avery Clifford – United States Marine Corps (1942–1945)
Lt. Louise McNeill-Davis – United States Army Nurse Corps (1942–1945)
Sergeant Louis "Louie" Gardenella- United States Marine Corps (1942-1945)
Sergeant Roger Hampton- United States Army (1942-1946)
Private Joe Harris-United States Army (1943–1946)
Private First Class Joe Hexter- United States Marine Corps (1942-1945)
Private Brad Holmes-United States Army (1944–1945)
Sergeant Ira Husker – United States Army (1942–1946)
Staff Sergeant Alfonse Mangano-United States Marine Corps, Fighter Pilot (1942-1946)
Corporal Narciso "Cheeso" Massaconi-United States Army, 1942–1945
Radioman 1st Class Bob Mercurio – United States Navy (1942–1946)
2nd Lieutenant Regina Moskowitz-United States Army Nurse Corps (1944 – 1946)
Private Don Palmer- United States Army (1942-1945)
Corporal Burt Paxton- United States Army (1942-1945)

11

Corporal Jimmy Rapp – United States Army (1942–1946)

Corporal Charlie Rickes- United States Marine Corps (1943-1946)

Electrician's Mate Dave Rugg-United States Navy (1943-1947)

Private Jim Semfy-U.S. Marine Corps (1942-1946)

Sergeant Gerry Stellman- United States Army (1943–1946)

Sergeant Irving "Irv" Tallon-United States Army (1943–1946)

Lieutenant Myron Walden-United States Army Air Corps (1942–1945)

Voices of the Greatest Generation

They were farm boys, city kids, sons, brothers, and fathers. Some lied about their age just to serve, while others were called upon before they even lived a full life of their own. From the beaches of Normandy to the dense jungles of the Pacific, they fought against tyranny with a resolve that would come to define them as the Greatest Generation.

Through these pages, you will hear their stories, in their own words, the harrowing accounts of battle, the camaraderie that sustained them, and the scars, both seen and unseen, that they carried long after the guns fell silent.

The veterans who served during World War II all shared common themes of sacrifice, resilience, the horrors of war, and camaraderie that they still feel to this day. The courage that these men and women exhibited is only matched by their humility and unwavering love for their country.

World War II was a war that was fought on not just one battlefield, but on three different fronts, The South Pacific, Europe, and North Africa. Each presented unimaginable challenges and the conditions they endured were as much an enemy as the Axis forces they faced.

The South Pacific: The Jungle's Relentless Grip

For those sent to the Pacific, the war was a constant fight against both a determined enemy and an unrelenting environment. The thick, humid jungles of islands like Guadalcanal, Peleliu, and Iwo Jima offered little respite.

On Iwo Jima, where volcanic ash replaced solid ground, movement was a slow, grueling task. **Private Charlie Rickes** described the chaos of landing on the black sand beaches: *"When we landed, we had to climb up these giant black sand dunes... Japs were firing at us left and right. I jumped on my belly and crawled about 25*

13

feet next to a soldier… he was a guy from my hometown… shot right through the forehead. God damn war sucks."

The Japanese laid in fortified caves and dug tunnels turning every hill into a killing field, and the battles often devolved into hand-to-hand combat.

Europe: The Frozen Hell and Fierce Resistance

The European Theater presented its own relentless hardships. The winter of 1944-45 was one of the harshest in history, and the soldiers who fought in the Battle of the Bulge faced brutal subzero temperatures with little more than their standard-issue wool uniforms. The relentless artillery fire and surprise attacks from German forces made every moment a fight for survival.

For those who landed on the beaches of Normandy, like **Private Don Palmer**, the conditions were no better. *"The morning of June 5th, we were supposed to go,"* Palmer remembered. *"But the weather was terrible—too foggy and overcast. Our Air Force planes couldn't fly, and the Navy had to postpone."*

The delay did little to calm their nerves. By dawn on June 6th, D-Day, the invasion began in earnest. Private Palmer's unit made their way across the channel toward the beaches of Normandy. *"I thank the good Lord every day that I wasn't in the first few waves,"* he said, visibly emotional. *"Jesus, they got massacred. The guys in those first waves... they didn't stand a chance."*

Further into Europe, soldiers encountered horrors beyond the battlefield. **Corporal Burt Paxton**, among the first to witness the liberation of the concentration camp Buchenwald, described the sights that haunted him forever. *"We were greeted by 10,000-15,000 men,"* he said, his voice heavy with emotion. *"The only word I can use to describe the scene was horrifying. How thin they were, their bones were almost coming through their skin. When they smiled, it was like a smile of death."*

North Africa: The Desert's Merciless Grip

Before Europe and the Pacific, many soldiers were sent to the deserts of North Africa, where the brutal heat and endless sandstorms created a battlefield unlike any other.

The battles against German Field Marshal Erwin Rommel's Afrika Korps were a test of endurance. The supply lines were thin, and rations were meager. Soldiers often went for days with little to eat, sustaining themselves on stale bread and canned meat while the enemy, more familiar with desert warfare, launched surprise raids in the dead of night.

Sacrifice, Survival, and Brotherhood

Despite the harsh conditions, the soldiers of World War II fought with an unbreakable spirit. **Sergeant Louis Gardenella**, spoke of the deep bonds formed in battle. *"Your squad became your family. You shared everything, rations, cigarettes, letters from home. When someone didn't make it, it cut deep."*

The challenges they faced, whether in the snow-covered forests of Europe, the steaming jungles of the Pacific, or the scorching deserts of North Africa, tested their bodies and spirits. Many never returned home, and those who did carried the war with them for the rest of their lives.

Yet, despite all they endured, these men did not see themselves as heroes. They simply did what had to be done. And when they came home, they built lives, raised families, and helped forge the modern America we know today.

Their stories are not just history, they are reminders of the courage, sacrifice, and resilience that defined an entire generation. We owe them more than just our gratitude. We owe them our remembrance.

Private Jerry Anacosta

United States Army (1942–1945)

The Battle of the Bulge, fought from December 16, 1944, to January 25, 1945, was one of the most pivotal and brutal encounters of World War II. It marked the final major German offensive on the Western Front, as Hitler's forces launched a surprise attack through the Ardennes region of Belgium and Luxembourg. Freezing temperatures, dense forests, and heavy snow compounded the difficulty of the battle, which involved over 1.1 million soldiers, including 500,000 Americans. It was a test of endurance, strategy, and sheer will, where Allied forces eventually turned the tide and dealt a significant blow to German ambitions. Among the brave individuals who served during this historic battle was Private Jerry Anacosta.

Private Anacosta served in the United States Army from 1942 to 1945 as part of the 12th Army Group. Growing up in Boston's North End, he was deeply shaped by the close-knit Italian American community that defined the neighborhood. The North End, a bustling enclave of immigrant families, was known for its vibrant culture, deep sense of loyalty, and unwavering patriotism. Many young men from this community, inspired by their familial ties and dedication to their adopted country, enlisted or were drafted during World War II. Among them were influential Italian Americans who went on to make significant contributions both on the battlefield and back home, such as Medal of Honor recipient Sgt. John Basilone, who became a symbol of heroism during the Battle of Guadalcanal and Iwo Jima.

Private Anacosta exemplified this spirit of sacrifice and determination. Drafted just after graduating high school, he trained at Fort Sampson, New York, before being deployed to the European Theater. While he fought in various campaigns, the

16

harrowing conditions of the Battle of the Bulge left an indelible mark on him.

"Jesus, it was cold… Christ sakes, you never could imagine THAT type of freezing cold," he said. *"Everything was cold— your food, water, coffee when you could get it. Initially, we weren't prepared for it, and they [the Germans] kicked our asses good… but we got 'em in the long run. They thought they were tough with all that crap about the master race—pure BS. We had the strength to not give in."*

The war's emotional toll weighed heavily on Private Anacosta, especially as two of his younger brothers were serving in the Pacific. Despite this burden, he displayed extraordinary bravery. He recounted a terrifying patrol: *"We were heading back right before sundown with about ten guys. We were wading through the snow and came to a road. Suddenly, we heard rumbling… a convoy of German tanks and trucks, about 3-4 hundred feet away. We dove about forty feet into the snow and buried ourselves. It felt like hours before they passed. If it had been daylight, we'd have been dead for sure."*

Private Anacosta also shared a poignant Christmas memory from the battlefield. During a skirmish, he saved a fellow soldier from a German grenade, sustaining shrapnel wounds in his left leg—a heroic act that earned him a Purple Heart. Sent to a field hospital on December 23, he woke to the sound of nurses singing Christmas carols.

"I didn't even realize it was Christmas Eve until then," he recalled. *"They brought in a young kid who looked just like my little brother Johnny. He was shot up really bad. The nurse said he wouldn't make it through the night."* Fighting back tears, he shared the moment that haunted him for decades. *"That kid… he died a few hours later. They moved him to the tent for the dead. He was the only one in there… just a god damn kid. I asked the*

doctor if I could sit with him so he'd have somebody on Christmas Eve... even if he was dead. Jesus, he looked just like my brother. I'll never forget that."

Today, Private Anacosta remains sharp-witted and fiercely independent, living alone with the help of home health aides. As we ended our conversation, he chuckled, *"Not ready to go to one of those places yet... not giving up my god damn TV remote to anyone!"*

Thank you, Private Jerry Anacosta, for your extraordinary service and sacrifice. You stand as a proud representative of the Greatest Generation.

Petty Officer Bernie Auge

United States Navy (1942-1946)

Dr. Bernie Auge served his country in the United States Navy from 1942 to 1946 as a 2nd Class Petty Officer, most notably in the highly classified field of Naval Intelligence. At 101 years of age, he remains gracious, remarkably sharp, and a distinguished representative of The Greatest Generation, carrying himself with extreme humility, pride, and distinction.

Bernie Auge grew up in North Adams, Massachusetts, where he excelled in both academics and athletics. A standout athlete in football and baseball during high school, he also achieved regional acclaim as a speed skating champion, even skating in the historic Boston Garden. His athletic prowess earned him an offer of an athletic scholarship to Williams College. However, driven by a sense of duty and a desire for a broader education, he chose to attend Notre Dame University. His connection to Notre Dame remains strong to this day, as he continues to express his deep school spirit.

After just three months at Notre Dame, Auge was drafted into the Navy as the nation prepared for World War II. He completed his basic training at Sampson Naval Training Base in New York. From there, he was sent to the University of Miami, Ohio, to study code and radio operations, essential skills for Naval Intelligence. His training led him to assignments in Washington, D.C., and ultimately to Cape Cod, Massachusetts.

In Cape Cod, Auge was stationed at the elite Naval Base at the Marconi Maritime Center in North Chatham—a ship-to-shore radio station built in 1914. The facility played a crucial role in intercepting and decoding enemy communications. *"We were sworn to secrecy under penalty of death—that's how top secret it was,"* Auge recalled. *"We never talked to anyone about what we were doing, not even my wife, until 20 years after the war."*

Auge's work as part of Naval Intelligence had a significant impact on the war effort. His station was one of five key listening posts tasked with intercepting German U-boat transmissions and pinpointing their locations in the Atlantic Ocean. The other stations were located in Greenland (The Rock), Charleston, South Carolina, Brazil, and Washington, D.C.

German U-boats posed a constant threat to American supply ships, sinking them with alarming regularity. Auge's intelligence group was instrumental in disrupting these attacks by providing critical information that helped Allied forces track and neutralize the submarines. Their work ensured the safe passage of supply convoys, a lifeline for both the military and the home front.

A pivotal moment in the intelligence war occurred when the Navy captured a German U-boat in a surprise attack. Typically, when captured, the German crew would scuttle their vessel by opening a valve to sink the ship. However, in this instance, the Americans managed to seize control before the crew could act.

"Nobody knew we had captured it—not even the Germans," Auge explained. *"They had it categorized as 'lost at sea.' We brought the boat and crew to Bermuda and were able to obtain their radio communication code books."* The codebooks provided a critical breakthrough, allowing Allied forces to decode enemy transmissions and further weaken the German U-boat threat.

The seriousness of their mission was evident in the day-to-day operations at the listening station. *"Everyone paid attention to detail and was quite serious. It was quite the duty—we could never leave our stations, and meals were brought to us there,"*

Auge recalled. The secrecy and intensity of the work created a unique bond among the sailors, though they were forbidden from discussing their duties.

Despite the secretive nature of his service, Auge reflected on his time in the Navy with pride. *"It was a different feeling. We were proud of the work we did and what we accomplished, but we just never talked about it. That was the main thing."* Remarkably, even at 101 years of age, Auge can still read Morse code—a skill he learned during his time in the Navy.

Following his military service, Auge pursued a successful career in dentistry, establishing a dental practice that served his community for 36 years. His commitment to service extended beyond his professional life. He married his high school sweetheart, Eleanor, and they shared a remarkable 78 years of marriage until her passing. Auge remains the proud patriarch of a large family, including 12 great-grandchildren and 44 great-great-grandchildren.

"I've had a pretty good life," Auge said. *"God gave me such a beautiful wife."* His humility and gratitude for life's blessings are evident in every word he speaks.

Dr. Bernie Auge's story is one of service, sacrifice, and dedication. His work in Naval Intelligence played a crucial role in altering the course of World War II, ensuring the safety of countless American lives. His post-war contributions to his community and his enduring legacy as a family man further highlight the values that define The Greatest Generation.

Dr. Bernie Auge, thank you for your service to our great country.

Private Charlie Barten

United States Marine Corps (1942–1945)

It is an honor to feature another member of the Greatest Generation, Private Charlie Barten who served his country with unwavering courage during World War II as a Marine in the South Pacific Theater from 1942 to 1945. Now 95 years old, Charlie is still remarkably fit, enjoying long walks and tending to his garden in Bourne, Massachusetts, where he resides with Merideth, his wife of nearly 70 years. *"We met at a USO dance, and it was love at first sight,"* Merideth said proudly. With a wink, Charlie added, *"Yeah, she hooked me pretty good."*

Drafted shortly after the attack on Pearl Harbor, Charlie was sent to Fort Sampson, New York, before joining the 3rd Marine Division. His first major operation was the invasion of Bougainville Island. Recalling those days, he spoke highly of the Seabees. *"The Seabees were worth their weight in gold. They built airstrips so our planes could land, and they saved us,"* he said. Their mission focused on the Japanese base at Rabaul, which he described as an *"unbelievably vicious enemy."* The psychological toll of the war was immense. *"At night, they preyed on our shell-shocked guys, calling out their names, saying their girlfriends left them or their mothers didn't love them. Some guys snapped, and you tried to hold it together, but you were in the same boat."*

Charlie's memories of the holidays in the Pacific were bittersweet. *"I came from a close family. I had two brothers fighting in Europe. We tried to make it work, but it wasn't the same. We were too busy trying not to get shot."* He vividly remembered the cruelty of the Japanese soldiers. *"They were merciless. The stories of what they did to captured prisoners—the hot boxes, water torture, bamboo slivers under fingernails—were horrifying. At*

22

night was the worst; they would sneak into foxholes. Nobody I knew ever slept well. Our nerves were frayed. Even after all these years, I still have nightmares."

In addition to the invasion of Bougainville, Charlie fought at the Battle of Balikpapan in Borneo, just before the atomic bomb was dropped on Hiroshima. When asked about someone he admired during the war, he immediately named Major General A. H. Turnage, commander of the 3rd Marine Division. *"He was fearless, a brilliant tactician who was always two steps ahead of the enemy. He hated what the Japanese did to our prisoners and always checked in on his men. That meant a lot to us."*

Private Barten was awarded the Purple Heart for his service during the Bougainville campaign but humbly changed the subject when it came up. As he approaches his 96th birthday, his sense of humor remains intact. When asked what he wanted for his birthday, he quipped, *"97."*

As we finished our conversation, Charlie shook my hand, looked me in the eye, and offered a heartfelt reminder: *"Don't let them forget about us."* His words are a call to honor and remember the sacrifices of those who served.

Private Charlie Barten, thank you for your remarkable service to our country. Your story and legacy will not be forgotten.

Private First Class Sam Baxter

United States Marine Corps (1942–1945)

Private First Class Sam Baxter served in the United States Marine Corps during some of the most intense years of World War II. Born and raised in Little Rock, Arkansas, Sam enlisted at just 17, following the example of his three older brothers who also joined the service. Now 98 years old, he remains a vibrant storyteller, his hearty laugh and sharp wit undimmed by the passage of time.

Sam began his journey at Parris Island, South Carolina, the legendary training ground for the Marine Corps. Known for its grueling regimen and relentless drill instructors, Parris Island was designed to break recruits down and rebuild them into disciplined Marines. *"The yelling of our drill instructors was intense,"* Sam remembered. *"Morning, day, and night—except when they were eating, because their mouths were full!"* Coming from a quiet, church-going family, the cacophony of Marine boot camp was a shock. *"My mother was the kindest, sweetest woman in the world, and the only time I ever heard my dad raise his voice was at our stubborn mule when it wouldn't move!"* he said with a laugh. Despite the challenges, Sam adapted, finding strength in the camaraderie of his fellow recruits.

After completing basic training, Sam embarked on a long journey to the South Pacific aboard a military transport ship. These ships were crowded and uncomfortable, with soldiers packed into bunks stacked three or four high. Meals were basic, and the smell of diesel fuel mixed with the salty sea air could be nauseating.

Sam was stationed in the South Pacific Theater, including time on the Solomon Islands and Guam. The Solomon Islands campaign was a pivotal part of the war, where Allied forces engaged in fierce jungle warfare. On Bougainville, Sam encountered harsh

24

conditions: stifling humidity, dense jungles, and the constant threat of enemy ambushes. Supplies were often limited, and the tropical environment bred disease and exhaustion. *"It wasn't pretty, that's for sure,"* he said.

One of Sam's most powerful memories from the war occurred during a patrol in the jungles of Bougainville. *"I was sent out as a forward observer, gathering intelligence on the Japanese. I was about 50-60 feet ahead of my patrol when I came into a clearing and locked eyes with a Japanese soldier,"* he recalled. *"We just stared at each other—it felt like forever. Neither of us blinked. I was ready to raise my rifle, but something stopped me. We seemed to communicate without speaking. Slowly, we both stepped back into the jungle, never taking our eyes off each other. I often think about him, wondering what became of his life. We chose to spare each other that day."*

During his time in Guam, Sam encountered British soldiers rescued from a Japanese prison camp. Their condition left an indelible mark on him. *"They looked awful—so drawn and gaunt, just skin and bone. Some of them were pleading for food, almost crying,"* he remembered. Medical personnel had to deny their requests for food initially, fearing that their malnourished bodies couldn't handle solid food. *"That memory will never go away,"* Sam said, his voice heavy with emotion. The plight of prisoners under Japanese captivity was horrific, marked by starvation, disease, and brutality. British soldiers, alongside American and Australian POWs, endured some of the worst atrocities of the war.

Despite the hardships, Sam found moments of joy. One Christmas on Guam, he discovered that his older brother Gene was stationed nearby. *"I actually stole a jeep from the motor pool,"* he admitted with a chuckle. *"I later gave the guy a carton of cigarettes and a bottle of booze so he wouldn't squeal on me."* After tracking Gene down in a hospital ward, the brothers embraced, crying and laughing. *"He jumped out of bed like he was shot out of a cannon. That was the best Christmas present I ever had."*

Today, Sam lives in Mashpee, Massachusetts with his daughter and son-in-law, enjoying his three grandchildren and one great-granddaughter. His reflections on service are a reminder of the sacrifices and resilience of the greatest generation.

Private First Class Sam Baxter, thank you for your service to our great country.

1st Lieutenant Alfred Benjamin

United States Army Air Corps (1942-1945)

At 96 years old, Al Benjamin is truly something special. His extraordinary energy, vitality, and sense of humor are unmatched for a man his age. The retired 1st Lieutenant grew up in Dorchester, Massachusetts, and from a young age, he was fascinated by flying. *"I read every magazine I could find on flying and every book about learning to fly and how to navigate,"* he recalled. When the U.S. Army published training manuals on air navigation, Lt. Benjamin devoured them.

In July of 1942, the Army Air Corps needed aircrew members. Lt. Benjamin passed the Aviation Cadet Exam with ease and was sworn into the Army Air Corps Reserve. The influx of applicants was overwhelming, so the Army asked universities to train Aviation Students to supplement their education. He was sent to Canisius College in Buffalo before moving on to the Cadet Classification Center and then Preflight Ground School at Maxwell Field. After graduating, he completed his flight training in Avon Park, Florida, and underwent nine weeks of basic training in Atlantic City, New Jersey, where soldiers were housed in hotels.

His first assignment placed him in the legendary 8th Air Force, a unit that would become known as "The Mighty Eighth." The 8th Air Force played a pivotal role in strategic bombing campaigns over Nazi-occupied Europe, striking industrial and military targets to cripple the German war machine. It was also one of the most dangerous assignments in the war. The average life expectancy of a bomber crew was only 13 missions, yet some flew as many as 35. The average age of an aircrewman was 21 years old, and flight officers were typically just 23. The statistics were staggering, 35% of crewmen were seriously wounded, and 30% were killed.

Lt. Benjamin became a navigator on a B-17 Flying Fortress, a four-engine heavy bomber known for its rugged durability and ability to take heavy damage. His first mission had him transporting K-rations from Nebraska to an RAF base in northern London, but soon, he was flying combat missions deep into enemy territory.

When asked about being away from home during the holidays, Lt. Benjamin brushed it off. *"We were too busy to think about it... We flew 2-3 days on a mission and then had 2-3 days off."* He would go on to fly an astonishing 31 missions, but it was his 13th mission, *"the unlucky 13th"* that he would never forget.

On that fateful day, Lt. Benjamin and his crew had just dropped their bomb load on factories and munitions plants over Germany when their B-17 was hit by enemy fire. The plane began losing altitude rapidly, and the captain prepared the crew to jump. But Lt. Benjamin pleaded for more time exactly 77 more minutes before bailing out.

"I had calculated the precise time we needed to reach the nearest known American battle lines or, at the very least, friendly territory," he explained. When the captain asked why he was so adamant about waiting, Lt. Benjamin didn't hesitate. *"Sir, I'm Jewish. If we jump into German territory and I'm captured, I'll surely be killed."*

As the plane dropped to 10,000 feet, the bailout bell sounded, and the crew leaped from the burning aircraft. When Lt. Benjamin landed, he was immediately surrounded by armed men. At first, he thought they were Germans, but they were French Resistance

28

fighters. They shouted at him in French, mistaking him for a German airman. It was one of his fellow crew members who ran forward, shouting, *"He's an American! He's an American!"*

The Resistance took them to a nearby farmhouse, where they hid out, as Nazi patrols were still in the area. That night, the fighters transported them to a local hospital, where they received medical treatment and food.

The next morning, Lt. Benjamin and his crew woke up to an incredible sight—townspeople lined up outside their hospital windows, eager to see the American airmen who had fallen from the sky. Some brought small gifts, others simply shed tears of joy.

One encounter particularly moved Lt. Benjamin. An elderly woman visited him, speaking in French while an English nurse translated. She told him about her two older sons—both executed by the Nazis as Resistance fighters. Her youngest son had been in hiding, waiting for the Americans to liberate them. Hearing her story brought tears to Lt. Benjamin's eyes.

Lt. Benjamin's 31 missions took 17 months and 31 days. When asked if he had ever been afraid, he answered honestly. *"I was afraid thirty-one times."* But he added, *"I never had a bad dream and have no regrets from my service."*

1st Lieutenant Alfred Benjamin, thank you for your service to our great country.

Seaman Lionel "Red" Blanchard

United States Navy (1941-1952)

Lionel "Red" Blanchard, holds the remarkable distinction of serving his country in two major conflicts—World War II and the Korean War. At 92 years young, Red is an incredibly kind and warm individual who was a joy to interview alongside his lovely wife, Sherry. Born and raised in New Bedford, Massachusetts, Red joined the Navy during his senior year at Holy Mary High School, though he later returned to complete his degree. His naval career spanned 11 years, during which he served aboard four different ships. His first assignment was aboard the USS Duluth, a Cleveland-class light cruiser.

The USS Duluth played a crucial role in the Pacific theater during World War II. Designed for speed and armed with impressive anti-aircraft capabilities, the Duluth was tasked with providing protection for Allied forces from both air and sea threats. Red's time aboard the Duluth was not without peril. He vividly recalled being caught in a typhoon in the South Pacific, where blinding rain and gale-force winds pushed the ship to a dangerous 59-degree roll. Tragically, the storm split the ship's bow in two, claiming the lives of 17 crew members. ***"We had to be towed into Okinawa for repairs,"*** he shared, his voice tinged with sadness.

Red's most harrowing memory aboard the Duluth involved surviving an attack by Japanese kamikaze pilots. These pilots, trained for deliberate, suicidal crashes, posed a relentless threat to Allied ships. ***"The kamikazes dropped out of the sky and skimmed just above the water,"*** Red recalled. ***"One Japanese***

30

pilot came bearing right down on us. He either got hit by shrapnel or lost his nerve because he veered off at the last second and crashed into the ocean. I hit the deck. You can't dig a foxhole on steel," he said with a mix of humor and gravity. The intensity of these attacks tested the nerves and resolve of every sailor aboard.

Red's journey continued on the USS Saturn, a supply ship that transported crucial goods to bases in Iceland, the Caribbean, Trinidad, and Rio de Janeiro. During his station in Iceland, he served as a lookout atop the ship's mast, searching for icebergs in the freezing, turbulent waters. *"It was bitterly cold, and the ship swayed constantly. I had eaten sausages earlier, and, well, let's just say they didn't stay down for long!"* he chuckled. He also spoke fondly of a mentor named Svenson, a 32-year "Mustang" who worked his way up the Navy ranks. *"He always had your back—a real crewman's guy. I really respected him."*

In Korea, Red faced further danger while stationed aboard a ship where an explosion tore through seven decks during welding work below. Tasked with inspecting the damage, he donned a breathing mask and descended into the smoky chaos. *"The first man I found was a guard at the ammunition locker,"* he recalled solemnly. *"All he had left on was his gun belt. The explosion had consumed him."*

Despite the hardships, Red found lighter moments during his service. As the manager of the ship's "Geedunk" (an ice cream stand), he fondly remembered delivering 20 gallons of ice cream to a downed pilot rescued by a destroyer. *"I joked with him, 'Hey, you made me miss my movie,' and he just grinned and said, 'Too bad!'"* Another favorite memory was his time aboard the USS Lejeune, ferrying troops overseas on the "Magic Carpet Route." To combat seasickness among the soldiers, Red handed out saltines— a simple gesture he took pride in, as none of his charges ever became ill under his watch.

Red's service was defined by duty, resilience, and camaraderie. Whether braving typhoons, fending off kamikazes, or savoring moments of laughter with his shipmates, his stories encapsulate the spirit of a generation that faced extraordinary challenges with courage and determination. ***"You don't experience anything; you go because you're told to go,"*** he reflected. ***"I wasn't a troublemaker. I followed orders and did what I was told."***

Seaman Lionel "Red" Blanchard, thank you for your unwavering service to our nation.

Private Avery Clifford

United States Marine Corps (1943–1946)

Private Avery Clifford served his country with unparalleled courage during World War II as a Marine in the South Pacific Theater from 1943 to 1946. Born in Baton Rouge, Louisiana, he grew up hunting, fishing, and wrestling alligators alongside his father and grandfather. By the time he arrived at Parris Island, South Carolina, for basic training, he was no stranger to tough challenges. *"Boot camp wasn't a big deal. I'd been dealing with 14-foot gators in the dark—drill instructors yelling in your face didn't faze me,"* he said with a chuckle.

Private Clifford's wartime experience would take him to one of the most grueling battles of the Pacific—Peleliu. The mood among the troops as they approached the beach in Landing Ship Tanks (LSTs) was somber. *"The worst part was seeing everything happen right in front of us. Our guys were getting mowed down on the beach like dominoes. It was a slaughter. My heart was in my toes,"* he recalled. Once the ramp dropped, the reality of combat was immediate and unforgiving. *"Guys took one or two steps off the LST and got picked off. There was no cover—just bullets, blood, and chaos."*

The lack of leadership during the landing left a deep impression on Private Clifford. *"Our sergeant froze, didn't know what the hell to do. He crouched behind a dead soldier while the rest of us were getting our asses handed to us. His cowardice cost us a lot of men. It's something I'll never forgive,"* he said, his voice tinged with anger. The memories of that day were visceral. *"The smell of gunfire and blood—it's distinct. If you've been in combat, you know it. It never leaves you."*

Private Clifford recounted the harrowing reality of the terrain at Peleliu, describing the lack of preparation and the devastating cost. *"There was zero cover, and we weren't prepared for the jagged coral terrain. It was like they were sending us straight to the slaughterhouse. We took that beach through sheer willpower and determination. The Japanese didn't have the heart we had, but it came at a hell of a price."*

The emotional toll of war was immense, and Private Clifford found ways to cope by staying detached from his fellow Marines. *"I didn't get close to anybody. If they got killed, I'd be thinking about them, and that can get you killed in combat. You just focused on surviving and doing your job."*

Reflecting on the holidays during his service, Private Clifford's response was raw and honest. *"There weren't any holidays. The only celebration I had was on the troop ship home. They gave us a decent meal and a sheet to sing Christmas songs. I ate, but after what we'd been through, the last thing I wanted to do was sing 'Jingle Bells.'"*

As he approaches his 100th birthday, Private Clifford expressed a mix of pride and concern. *"What we had back then, as Americans, was something special. We were hungry—hungry to fight, hungry to win. Now? I don't see anything close to it. We've gotten soft—too god damn soft."*

Private Avery Clifford, thank you for your extraordinary service to our great country. Your story is a powerful reminder of the sacrifices made by the Greatest Generation.

Sergeant Louis "Louie" Gardenella

United States Marine Corps (1942-1945)

Louis "Louie" Gardenella served his country in the United States Marine Corps as a Sergeant from 1942 to 1945 in the South Pacific Theater. At 97 years old, he remains remarkably fit, with thick, leather-like hands that bear the marks of a hard life well-lived. His memory is sharp as ever, reflecting the resilience and grit of a man who lived through one of history's most challenging periods. Louie is, without question, a throwback to the "good old days."

Born in the Bronx, NY, Louie shared vivid memories of his early childhood. *"I had seven siblings. We were close—nobody came between us. My two older sisters were like mothers to me. I was the youngest. My older brothers were pretty tough, but everyone feared my sisters, Anna and Maria. One time, some kids were picking on me, and Anna raced down the stairs and chased them down the street in her high heels, cursing them all the way. Lucky for those punks she didn't catch them!"* he recalled with a hearty belly laugh.

Louie's military journey began at Parris Island, South Carolina, where he underwent basic training. *"Everybody thought the Drill Sergeant was tough,"* he remembered. *"Ah! Tough was in my neighborhood. I fought just about every day when I was a teenager."*

After several minor deployments across the South Pacific, Sergeant Gardenella and his unit were sent to New Guinea, where the 3rd Marine Division launched two major offensives on Bougainville Island in the Northern Solomon Islands. It was here

35

that he witnessed the crucial role played by the Navy Construction Battalions, more commonly known as the Seabees.

"The Seabees—we owe a great deal to them," Louie said with reverence. *"They came in and built the critical airstrips so our planes could land. The Japs had field artillery positioned on higher ground, and they wreaked havoc on those airstrips. The Seabees worked under constant fire, but they never stopped. They were brave SOBs."*

The Seabees were renowned for their ability to perform construction tasks in war zones. Their motto, *"We Build, We Fight,"* perfectly encapsulates their dual role as both builders and defenders. These men were skilled tradesmen—carpenters, electricians, mechanics—who rapidly constructed essential infrastructure like airfields, roads, and bridges in some of the most inhospitable and dangerous environments of the war. On Bougainville, their work was pivotal to the success of the Marine Corps' operations.

"They dug in pretty good," Louie continued. *"The Japs had a ridge that overlooked the beach. We called it 'Hellzapoppin' Ridge.' It was brutal. Without those airstrips, we would've been sitting ducks. The Seabees made sure we had what we needed to keep going."*

When asked about entertainment or downtime during his deployment, Louie gave a curt reply: *"No! We sure as hell could've used it, but we were fighting. The Japs always tried to get into your head. I put the fear of God into my guys by telling them, 'YOU NEVER GET OUT OF YOUR GODDAMN FOXHOLE AT NIGHT!' It could mean instant death by sniper or a Jap knife."*

He recounted a harrowing lesson from early in the campaign. *"In the first wave, a guy got out of his foxhole to urinate, and a sniper got him. After that, I told my guys, 'You go to the bathroom in your foxhole or your helmet. I don't care if it's number one or number two, but you don't leave your foxhole.'"*

When the conversation turned to holidays spent overseas, Sergeant Gardenella's demeanor shifted. He grew quiet for a moment before speaking. *"Not going to lie to you, kid. It was pretty damn tough. I came from a close family. I had two brothers over in Europe fighting. You knew it was there, but you had to work through it. It was really tough when the Japs piped in Christmas music over their loudspeakers. Some guys just couldn't take it."*

I asked Louie if there was anything he considered truly special about his time in the service. He grew visibly emotional as he answered. *"Being with my guys—the camaraderie we shared in the middle of death and carnage. That was special. You really relied on each other. You had to, or you were dead."*

Louie spoke fondly of the reunions he used to attend with his fellow Marines. *"Used to go to reunions, but there are very few of us left now,"* he said, his voice catching with emotion.

Today, Sergeant Gardenella lives in Bourne, Massachusetts. A widower since his beloved wife Isabella passed away two years ago, he continues to carry the memories of his service with pride and dignity.

Sergeant Louis Gardenella, thank you for your service to our great country. The legacy of the Seabees and the Marines you fought alongside endures in the freedom we enjoy today.

Sergeant Roger Hampton

U.S. Army (1942-1946)

Sergeant Roger Hampton served his country with distinction in the United States Army from 1942 to 1946. At 100 years of age, his remarkable story offers a vivid glimpse into the sacrifices and resilience of the Greatest Generation.

Born in Topeka, Kansas, Roger enlisted at the age of 18. However, in a testament to his determination, he recalls, *"I was 16 and a half at the time but lied about my age so I could get in after the Japanese bombed us at Pearl Harbor."* His patriotism and drive to serve led him to basic training at Fort Dix, New Jersey, before heading to Cape Cod, Massachusetts. There, he and 1,200 other soldiers underwent rigorous amphibious assault and landing operations on Washburn Island as part of an eight-week assignment at Otis Air Force Base (now Joint Base Cape Cod). He was soon assigned to the legendary 2nd Armored Division, famously known as "Hell on Wheels."

"Our nickname was 'Hell on Wheels,' and that sure as heck described us to a 'T,'" he recalled with pride. The division's fierce reputation would follow them through some of the most intense battles of World War II.

Sgt. Hampton's first mission in Northern France left an indelible mark on him.

"We had all this preparation and strategy drilled into us. Couldn't sleep the night before—hell, I don't think anybody could. They dropped us at night—pitch black and damn windy. Almost too windy. We missed our drop spot and got blown all over. We were scattered all over Northern France."

38

The disorientation from the drop was only the beginning of the chaos. Roger and a small group of soldiers, about eight in total, regrouped at an abandoned farmhouse, seeking shelter until dawn. As they navigated through the dense woods, they stumbled upon a haunting scene—three paratroopers hanging from a pine tree, each shot to death.

"That shook me up pretty bad because it was my first experience with death," he shared solemnly.

The group spent the next ten hours dodging German patrols, including the feared SS forces, before finally reuniting with their regiment. *"Let me tell you, boy, did we dodge the Germans. We saw a ton of them. God, we were so happy when we finally caught up with our troops."*

Sgt. Hampton saw heavy action on numerous occasions, including an intense 11-day stretch of continuous combat.

"Our motto was: eat when you can and survive every day. It was pretty simple if you wanted to stay alive," he explained. *"They had us almost pinned down once, but our guys had a determination and will to win like none other. Damn proud of the way my men fought."*

He shared that the holidays during wartime were especially tough. His first Christmas overseas, at the age of 18, was a stark contrast to the festive celebrations he had known at home.

"I was definitely a changed person from when I went in. I enlisted as a young boy and became a hardened man. We didn't have time to think about Christmas or the holidays. We were trying to stay alive every day. I was either in a foxhole or hiding behind a tree from gunfire. I don't remember anything about the holidays."

When asked about his fondest memory from his service, Sgt. Hampton's response was somber.

"When you're in combat, there aren't too many memories you're fond of. All I know is we lost a great many men, and I still sleep like hell thinking about the guys we lost—now more than ever."

Despite the hardships and loss, Sgt. Hampton's bravery and dedication did not go unrecognized. He was awarded the Bronze Star for his service and courage.

As we reflect on Sgt. Roger Hampton's story, we are reminded of the sacrifices made by those who served in the most trying of times. His resilience, courage, and unwavering commitment to his fellow soldiers exemplify the very best of what it means to serve.

Sergeant Roger Hampton, we honor your service and sacrifice. Thank you for your invaluable contribution to our great country.

Private Joe Harris

United States Army (1943–1946)

Private Joe Harris served his country with humility and resilience as a member of General George Patton's legendary 3rd Army during World War II. At 98 years old, Joe's sharp wit and warm country charm are as vibrant as ever.

Born in Hackleburg, Alabama, Joe grew up as the eldest of four children in a hardworking farming family. *"All we knew was hard work. My daddy used to say it was good for the soul, but at 4 a.m., when my brother and I had to milk 20 cows, my soul didn't feel too good,"* he recalled with a laugh.

After basic training at Camp Shelby in Hattiesburg, Mississippi, where the sweltering summer heat made the experience memorable, Joe received additional training at Camp Bowie in Texas. From there, he boarded a troop ship alongside 500 soldiers bound for Europe. *"As far as my eyes could see, to the left and right, there were ships. So many ships,"* he remembered.

Landing in France just before winter, Joe was assigned to small arms repair near the front lines in Belgium. The bitter cold and relentless German offensive marked some of the darkest days of his service. *"We saw a lot of wounded and dead. I got assigned to drive a truck, transporting them to medical tents for surgery or to the morgue. The morgue was just a tent—nothing else,"* he said, his voice quiet with reflection.

Although Joe never formally met General Patton, he had a few close encounters with the larger-than-life leader. *"I came about three feet from him a couple of times. His temper was legendary. One day, he caught a private without his helmet—God help that poor soldier. He screamed so hard I thought he'd have a stroke.*

41

His aides had to pull him away; otherwise, I think he would've pulled his gun," Joe recounted.

Holidays during wartime were bittersweet. *"They gave us a hot meal, but the guys on the front lines didn't get one because of the heavy German fire. The first time I heard Christmas music, it really got to me. I was coming out of the medical tent after dropping off some wounded, and I heard 'Silent Night.' Then I walked past the morgue and stopped to look in. It hit me hard—those soldiers would never have another Christmas,"* he said softly.

Reflecting on his service, Joe shared, *"It was such an honor to serve my country, especially at a young age. On the train home, every time a buddy from my unit got off at a stop, it felt like saying goodbye to a brother—we were brothers in some ways. I wasn't a hero, not like the guys on the front lines or the wounded I saw. I saw what war does to a person."*

Today, Joe enjoys visits from friends, his children, and grandchildren, who keep his home filled with laughter and love.

Private Joe Harris, thank you for your extraordinary service to our great country. Your story is a poignant reminder of the sacrifices made by so many in the fight for freedom.

Private First Class Joe Hexter

United States Marine Corps (1942-1945)

Joe Hexter served as a Private First Class in the United States Marine Corps, 1st Marine Division, during World War II from 1942 to 1945. Born and raised in Nashville, Tennessee, he enlisted at the age of 18, driven by a sense of duty to his country during a time of global conflict.

Private Hexter began his military journey at the Marine Corps Recruit Depot, Parris Island, North Carolina. Reflecting on his time in basic training, he said, *"It was ok... other than getting you in shape, they really focused on discipline and paying attention to detail... they were on you constantly... and you paid the price individually or everyone paid the price if you screwed up."* His words capture the relentless focus on teamwork, accountability, and readiness that defined Marine Corps training during the war.

Joe Hexter saw extensive combat in the South Pacific Theater, particularly on the islands of Palau. The conditions were grueling, and the dangers were ever-present. *"We spent a lot of time in the jungles... the fighting was intense... hated nighttime... still do... absolute worst... didn't know if a enemy soldier was 2-3 feet from your foxhole... couldn't see a damn thing,"* he recalled.

He described the challenges of navigating dense jungles under constant threat from an unseen enemy. *"We had a couple of service dogs, and they were the best... the Japanese were afraid of them. It was not a smart thing to get out of your foxhole in pitch black in the middle of the night... the enemy was everywhere,"* he said.

One harrowing memory stood out. *"We had a Marine go into the mouth of a cave to take a bathroom break, and a Japanese soldier almost ran him through with his sword... he'd been hiding in there. It scared the living hell out of the Marine. Don't think he was ever the same after... then again, I don't think any of us were."*

When asked if he experienced holidays during his time in the Pacific, Private Hexter shook his head. *"Never... we were in the jungles. You might have said 'Merry Christmas' to the guy in your foxhole or to a few guys, but we didn't do any celebrating... no special meals... no nothing. Where we were, we didn't expect any visits from Santa Claus,"* he remembered, his voice tinged with a mix of humor and sorrow.

Private Hexter's wartime experiences left a lasting impact on him, shaping his life long after he returned home. When asked if he had any regrets, he paused before answering. *"When I was younger,"* he said, taking a deep breath, *"I used to say the only regret I had was that I didn't kill more enemy soldiers."*

His expression grew solemn as he continued, *"I look back now on my life and realize that wasn't the best answer. War screwed me up pretty good... I wasn't the same person when I came home... totally different. I look at my grandkids now, and I was dodging bullets at their age... saw a ton of death... injuries... all caused by goddamn war. Those memories never go away."*

Tears filled his eyes as he spoke of his evolving perspective. *"I look back now, and the enemy was doing what they were ordered to do, just like we were. It really haunts me... I killed young men who happened to be the enemy. I had so much hate toward them. It slowly has disappeared over the years. A bunch of Marines I served with went back many years ago... wish I had."* His voice trailed off, and he shook his head, lost in thought.

Today, Joe Hexter resides in Mashpee, Massachusetts, where he enjoys spending time with his grandchildren and great-grandchildren. Despite the lasting scars of war, he finds solace in family and the peace that eluded him during his youth.

Private Joe Hexter, thank you for your service to our great country.

Private Brad Holmes

United States Army (1944–1945)

Private Brad Holmes is a proud member of The Greatest Generation, whose story of resilience and sacrifice during World War II is nothing short of extraordinary. At 96 years old, Brad still possesses a strong handshake, a sharp memory, and an unshakable sense of pride for his service to his country. Captured during the Battle of the Bulge, Brad spent over a year as a prisoner of war in a German camp, enduring unimaginable hardships with unwavering courage.

Born and raised in Everett, Massachusetts, Brad was sent to Fort Devens for basic training, followed by several other training camps, including Croft in South Carolina, Shelby in Mississippi, and Atterbury in Indiana. He finally boarded the Queen Elizabeth II in New York to join hundreds of other soldiers heading to Torrington, England. After crossing the English Channel, his unit was deployed to Belgium and France. *"The snow was thick on the ground, and it was tough to dig a foxhole,"* he remembered.

Deployed near the Ardennes Forest, Brad and his unit faced the harsh realities of combat during the Battle of the Bulge. *"We'd been on the front lines for about two weeks, and it was quiet— then all hell broke loose. There was gunfire, bullets flying everywhere. I dove for cover behind a dead cow. Those 'Screaming Mimis'—the 88s—were terrifying,"* he recounted.

Captured by German forces around Christmas and New Year's, Brad and his fellow soldiers were ordered to unload and destroy their weapons. *"They searched us for anything valuable, lined us*

46

up on the road, and marched us ten miles to a boxcar. They were rough with us. At night, the RAF bombed the area, not knowing we were in those boxcars. I told the guys we'd be fine unless we got a direct hit," he said. Tragically, some prisoners panicked and attempted to escape in the dark.

Brad was eventually taken to a prison camp housing thousands of soldiers from various nationalities. The Germans sought volunteers for work details, and Brad volunteered, hoping for better conditions. *"We marched 15 miles, were given burlap bags to fill with straw, and that's what we slept on,"* he said.

Life as a POW was grueling. Brad worked in a furniture factory, carrying logs all day in the woods. *"It was a killer. They fed us nothing but turnip and carrot soup—it was terrible. They locked us on the second floor of the factory at night and took our shoes to prevent escapes. Sometimes you were so weak you had to crawl up the stairs on your hands and knees,"* he remembered. Holidays were a distant memory. *"Oh yeah, they were non-existent."*

As the Russian forces advanced, Brad and the other POWs were moved closer to the Czech border. One day, liberation came in an unforgettable moment. *"We heard planes, and I saw one of ours swoop down. I waved to him, and he actually waved back before strafing a line of German trucks. BOOM! They were gone. The guards just disappeared—they knew it was over."*

Even after enduring such hardships, Brad never lost his humanity. *"I saw a German woman pulling a cart with her belongings and a baby. She was struggling, so I helped her. She offered me the baby's milk, but I told her to keep it. Then I saw a jeep with a sign that said 'Press Corps,' and I knew the Americans weren't far behind."*

Brad's journey home was swift. *"They put us on a C-47 and flew us to Camp Lucky Strike in France, where we got new clothes, and they burned our old ones. We'd been wearing the same ones for months. They gave us big canisters of eggnog. I came home on a Navy ship and slept in a hammock,"* he said with a smile.

Reflecting on his service, Brad humbly said, *"I did the best job I could with what I was assigned to do. I was proud to serve my country."*

Private Brad Holmes, thank you for your incredible service and sacrifice for our great nation. Your story will never be forgotten.

Sergeant Ira Husker

U.S. Army (1942-1946)

Ira Husker served his country with honor as a Sergeant in the
United States Army from 1942 to 1946. Born in Des Moines, Iowa,
he was the eldest of three siblings born to Ira Sr. and Katy Ann
Husker. His family's patriotic commitment ran deep. *"My middle
brother Jeb was a forward observer in Patton's 3rd Army, and my
youngest brother Toby was a seaman in the Navy in the South
Pacific Theater,"* he recalled with pride.

Enlisting at the age of seventeen, Ira completed his basic training
at Fort Sill, Oklahoma. *"I worked on a farm, so the physical stuff
didn't bother me none. Gettin' up at 5 a.m. was like sleeping in
for me—I used to get up at 3 a.m. to do my chores,"* he said with
a chuckle.

Sergeant Husker shipped out along with about a thousand other
soldiers on a troop ship bound for England. The journey was
rough. *"We had some rough seas on the way over. Lots of guys
were blowing chow all over the place—medics on board gave out
a lot of saltines,"* he remembered with a laugh. But even
lighthearted moments couldn't mask the emotional strain. *"My first
holiday away from home was in England. It was tough—
seventeen, being away from your family. They did what they
could with a good meal and some Christmas music, but
sometimes the music made it worse. The only thing that got me
by was knowing all the other guys were in the same boat."*

Among his most vivid memories was his role in Operation Market
Garden, one of the largest airborne operations of World War II. On
September 14, 1944, Sergeant Husker and his comrades parachuted
into Holland. *"We were to be dropped way behind the lines of
occupied Holland where the Germans had a stronghold,"* he
explained. *"Dutch resistance was wreaking havoc with German*

forces. If we defeated them, our tanks would be able to go into Germany without much resistance."

The jump itself was harrowing. *"I remember the roar of the engines and the cold wind rushing past as we jumped from the plane. The sky was filled with parachutes—you could hear gunfire in the distance. You're coming down fast, and all you can think about is getting to the ground without getting shot."* Landing safely was only the beginning of their challenges.

"Kept seeing farmhouses and houses with orange ribbons on them—a symbol of Dutch resistance," he said. As they advanced into Eindhoven, the reception they received was unforgettable. *"It was a madhouse! People lined the streets and were cheering like crazy. Men slapped us on the backs and shook our hands, women hugged and kissed us, and priests were giving out cigars."*

However, even in the midst of celebration, sobering sights lingered. *"One thing that still sticks in my memory is when they brought out 8 to 10 women, stripped them down to their undergarments, and other women started shaving their heads. We sat there dumbfounded. I asked one of the men next to me why they were doing this. He spoke little English but said, 'They sleep with Germans,' then he spit on the ground. I'll never forget that sight—the women screaming and crying."*

Later, as his unit moved out of the town, they encountered a young woman with a shaved head cradling a baby. *"She couldn't have been more than twenty. Several of us threw her our ration packs. You couldn't help but feel for her,"* he recalled, his voice heavy with emotion.

Reflecting on his service, Sergeant Husker expressed enduring pride. *"I'm very proud of what me and my fellow soldiers did. Think constantly of the guys we lost—guys that got shot and had a chance to make it but didn't. I really never got close to anyone because I couldn't stand the pain of losing them as a friend."*

50

At ninety-nine years of age, Sergeant Ira Husker resides in Mashpee, Massachusetts, receiving frequent visits from his four children. His life remains a testament to courage, resilience, and the enduring bonds of brotherhood formed on the battlefield.

Thank you, Sergeant Husker, for your service to our great country.

Hazel Jacobs

Women's United States Army Corps (1943-1945)

This veteran spotlight personifies the great American Dream. From humble beginnings in a small rural town in Louisiana to working in a top-secret capacity for the United States Army at the Pentagon in Washington, D.C., Hazel Jacobs' story is one of perseverance, courage, and patriotism. At 99 years young, Mrs. Jacobs remains as endearing and sharp as ever, her Southern charm and dignified presence shining through every word. From a young age, it was clear—she was absolutely fearless and ready to take on the world.

Born in Sulfur, Louisiana, Hazel lost her mother when she was just five years old. After graduating from high school in 1937, she went to work at a sugar refinery, where she filled one-pound boxes and ten-pound sacks of sugar. She also ran the switchboard a few nights a week, an experience that would later prove invaluable during her military service. One day at work, a conversation with a coworker changed her life's trajectory. *"My friend said she had just joined the Army,"* Hazel recalled. *"I said, 'You can't join— you're a woman!'"* Her friend explained that a new program had been created that allowed women to enlist. Inspired, Hazel enlisted in the Women's Army Corps (WAC) in March 1943 at the age of 23—a decision that was both bold and unprecedented for women of her time.

The creation of the Women's Army Corps (WAC) was part of a historic shift in World War II. Before the war, women had limited opportunities in the military, primarily serving as nurses. However, with the war raging and a shortage of manpower, the U.S. government realized the need for women in non-combat roles to free up men for active duty. In 1943, the WAC became an official part of the Army, allowing women to serve in uniform. By the end of the war, over 150,000 women had served in the WAC, making crucial contributions as clerks, cryptographers, mechanics, and switchboard operators—roles that directly supported military operations worldwide.

When Mrs. Jacobs first attempted to enlist, she encountered an unexpected obstacle: her weight. *"I saw a big sign outside the doctor's office,"* she recalled. *"It said, 'NO ENLISTEES UNDER 100 POUNDS.' Here I am, 5'1" and I've weighed 98 pounds my whole life—I was afraid they wouldn't take me!"* A friend gave her some quick advice: *"Eat a couple of bananas and drink lots of water."* Luckily, she made the cut and was accepted.

Basic training took place at Fort Oglethorpe, Georgia, for six weeks, after which she and a group of women were put on a train—destination unknown. *"The last thing I ever thought I'd see as I looked out the window was the White House and all the other buildings. I couldn't believe I was going to be working in Washington, D.C.,"* she said.

Because of her prior switchboard experience, Mrs. Jacobs was assigned to a classified position at the Pentagon, handling high-level, top-secret communications. She recalled her commanding officer, Colonel Becket, who took good care of the women under his leadership. *"We were called Becket's Girls, and he really looked out for us,"* she said.

Her job was nerve-wracking—monitoring and routing high-level military calls while ensuring absolute confidentiality. Security was tight. *"There was always a guard at the door, and we had to sign in and out whenever we worked,"* she remembered. Many women in this high-pressure role lasted only a year, but Hazel persevered.

While stationed in D.C., Mrs. Jacobs had the opportunity to rise through the ranks, eventually working in the Chief of Staff's Office, where she handled incoming and outgoing messages from every theater of war. She was also responsible for decoding top-secret messages and became skilled in using a Ditto Machine—an early document duplicator used for classified communications.

Among her most profound memories was seeing President Franklin D. Roosevelt in person during basic training at Fort Oglethorpe. *"I came out of the barracks, and the whole field was set up like a parade. I looked to my right, and there was President Roosevelt, sitting in his big black limousine with a blanket draped over his legs. I just couldn't believe it."*

However, her most solemn and deeply moving moment came in April 1945, following Roosevelt's death. Her WAC battalion was selected to march directly behind the President's funeral procession. Being a part of such a historic and deeply emotional event was something she never forgot.

Despite the wartime hardships, Mrs. Jacobs relished the opportunity to experience new places and friendships. *"I was so excited about being in Washington, D.C., that I never thought about missing home,"* she said. She fondly recalled her first trip to New York City, standing in awe in Times Square. Another cherished memory was seeing the legendary actress and singer Martha Raye perform at a USO show for the troops.

Mrs. Jacobs served honorably and, when asked about her time in the Army, responded without hesitation: *"I loved every minute of it."* Her military decorations and citations include, The American Theater Campaign Medal, Service Ribbon, and The Good Conduct Medal.

Mrs. Hazel Jacobs, thank you for your service to our great country.

Lieutenant Marc Jaffe

United States Marine Corps (1943-1946)

As I sat with this veteran, I was struck by his remarkable memory at age 98, as well as a defining trait of the Greatest Generation, humility. Lieutenant Marc Jaffe served in the United States Marine Corps from 1943 to 1946, seeing combat in some of the most grueling battles of World War II. Raised in Philadelphia, PA, he embarked on his military journey at the famed Parris Island for basic training, a rite of passage for every Marine. *"As we were arriving, there was a group of Marines marching by... they all yelled, 'You'll be sorry!'"* he recalled with a laugh.

Following basic training, he was selected for Officer Candidate School (OCS) at Quantico, Virginia, where he underwent rigorous leadership training before receiving his commission. After graduation, Lieutenant Jaffe was sent to Camp Lejeune, North Carolina, where he joined a replacement battalion and prepared for deployment to the Pacific.

Camp Lejeune, established in 1941, quickly became a critical training ground for Marines heading into battle during World War II. The base was designed to prepare troops for amphibious warfare, which was essential in the Pacific Theater. Recruits endured grueling physical conditioning, live-fire drills, and jungle warfare training, all in preparation for the brutal island-hopping campaigns against the Japanese. The training routine was relentless, designed to harden Marines both mentally and physically. Swamps, extreme heat, and the challenging coastal terrain mimicked the conditions they would face in the Pacific.

It was at Camp Lejeune that Lieutenant Jaffe trained in 81mm mortar operations, a role that required precision and quick decision-making under fire. The replacement battalion was constantly on high alert, as Marines trained with the expectation that they could be deployed at any moment. By the time Jaffe left Camp Lejeune, he was battle-ready, prepared for the unrelenting combat that awaited him in the Pacific.

After completing training, Lieutenant Jaffe was assigned as a forward observer in the 81mm mortar platoon, 2nd Battalion, 1st Marines. His first deployment took him to Cape Gloucester, New Britain Island, but it was Peleliu where he experienced his first major battle.

"We were in the first wave at Peleliu... had 26 members in the platoon that were either killed or wounded. I remember throwing myself on the sand—it seemed unreal, almost as though I was in a movie." The conditions were brutal. *"Each end of the beach was a high coral rock with enemy machine gun emplacements. The naval bombardment couldn't knock them out, so there was constant machine gun fire across the landing. We had to keep moving, scrambling to get into the trees and bushes."*

The chaos of the landing was unimaginable. *"There was so much confusion... had a Japanese tank attack us on the first day. The tank went right by me—I dove into a hole about two feet away."* He recalled how luck, training, and instinct helped him survive. As a Forward Observer, his job was to direct mortar fire, a critical role that allowed the Marines to suppress enemy positions. *"I was lucky to get off that beach."*

When asked about the camaraderie of his unit, Jaffe didn't hesitate. *"The most important thing I learned at Peleliu was the bond with the guys I served with. I wanted to be with them all the time. Comradeship is crucial in battle—whether it's Peleliu or any other fight."*

Following Peleliu, Lieutenant Jaffe and his unit were sent to Okinawa, the largest amphibious assault in the Pacific Theater and the bloodiest battle of the war in the Pacific. The battle, which lasted from April 1 to June 22, 1945, was a brutal fight that involved over 180,000 U.S. troops, including the 1st and 6th Marine Divisions and the 77th and 96th Infantry Divisions of the Army.

Okinawa was a fight to the death—the Japanese, knowing that losing the island meant an open path to Japan's mainland, fought with relentless ferocity. Over 110,000 Japanese soldiers were killed, and thousands more chose ritual suicide over surrender.

Lieutenant Jaffe recalled the brutal conditions Marines endured on the island. *"The rain was relentless… we were constantly drenched, living in trenches filled with mud. I had a poncho over my head at night, just trying to stay dry."* The terrain was treacherous, with dense forests and networked caves that the Japanese used for ambushes.

As the battle dragged on, Jaffe was given command of G Company, one of the battalion's three-line companies. *"The most important thing I learned was that you had to stay focused on the mission. You see someone go down who isn't going to get back up, but you have to keep moving."*

Lieutenant Jaffe also performed a heroic act of battlefield medical care. *"A Marine in my unit had been shot in the neck, and blood was flowing out of him. I acted quickly to stop the bleeding, keeping him alive until medics could take over."* Years later, at a unit reunion, the man introduced Lieutenant Jaffe to his wife. *"This is the man who saved my life."*

By the time Okinawa was secured, over 12,500 American troops had been killed and 38,000 wounded—the highest U.S. casualty rate of any Pacific battle. Many Marines, including Jaffe, fully expected that the next invasion would be mainland Japan—a battle that was estimated to result in over a million Allied casualties.

Then, in August 1945, the atomic bombs were dropped on Hiroshima and Nagasaki, leading to Japan's surrender. *"There was a huge sense of relief. We had been given the invasion plans for Japan—and now, we knew we wouldn't have to use them."*

As I prepared to leave, he pointed out a plaque on the wall—his Bronze Star commendation. As I read it, I saw it included the phrases, *"Coolness under pressure," "Good judgment in the line of fire,"* and *"Solid decisiveness in battle,"* all hallmarks of a great leader.

Lieutenant Jaffe's story is one of dedication, leadership, and unbreakable bonds. His courage under fire, ability to lead under extreme pressure, and commitment to his fellow Marines define the very best of the United States Marine Corps.

Lieutenant Marc Jaffe, thank you for your service to our great country.

Staff Sergeant Alfonse Mangano

United States Marine Corps (1942-1946)

It was an absolute honor to interview Staff Sergeant Alfonse Mangano, who served with distinction during World War II. Alfonse Mangano proudly served as a fighter pilot in the United States Marine Corps from 1942 to 1946, attached to the 3rd Marine Air Wing. Despite suffering from dementia, with his son by his side, the 97-year-old veteran vividly recalled his experiences from over 70 years ago.

Alfonse Mangano grew up in Brooklyn, NY, in a family with a proud tradition of military service—his father and two brothers were all Marines. His journey began at Parris Island, where he endured the rigorous boot camp. Transforming from a scrawny 140 pounds to a solid 170 pounds, Mangano's physical and mental resilience prepared him for the demanding role he would play in the war.

Staff Sergeant Mangano served in the Marine Air Corps, a division renowned for its role in sinking more Japanese aircraft carriers and battleships than any other unit in the South Pacific. His primary responsibility was conducting air patrol missions from Midway to Guam, scouting for enemy ships and submarines.

During our conversation, he shared a harrowing memory that still stirs deep emotions. *"We got our navigation mixed up somehow and flew over the sea near enemy territory. Japanese ships began firing on us, and they hit our fuselage. It was a major hit. We started losing fuel and had to hightail it out of there. We ended up losing all our fuel and glided for as long as we could, which wasn't very long. We crashed in the water. It was one goddamn scary feeling. My life jacket was caught on my seat, and I struggled to free myself. So much went through my mind as the plane started to sink. I finally freed myself and got to the surface gasping for air. All the crew was unharmed. Must have waited four to five hours in the water, though it seemed like fifty hours. Thought it was all over when we saw a shark fin circling us. Suddenly, we saw a boat off in the distance. We prayed to God it wasn't a Japanese ship. Thank goodness it was an American ship. They hauled us in and gave us dry clothes, smokes, and coffee. Jesus, I was so scared and shaking, my hands kept spilling the coffee."*

He continued with another chilling recollection: *"We were just so damn fortunate. We heard stories from captured Japanese about what they did when you were shot down and they plucked you out of the water. If you didn't tell them what they wanted, they tied your hands behind your back and tied your feet to an anchor, then threw you overboard,"* he recalled, cringing at the memory.

One of Staff Sergeant Mangano's most emotional recollections was witnessing a burial at sea. *"It is a very emotional experience. It broke my heart, seeing all those dead bodies in the bags— all young kids. The average age was nineteen, I think. It's something I'll never get out of my mind. Very, very tough."*

Reflecting on being away for the holidays, Mangano shared how letters from home brought him comfort. *"I got lots of letters. My mother was a write-a-holic and wrote her boys constantly. She always wanted to know how we were and what was going on with us. Of course, I didn't tell her everything as I didn't want her to worry."*

When asked about entertainment during his time overseas, Mangano chuckled as he recalled a memorable encounter. *"The shows were wonderful. I actually got a smooch from actress Rita Hayworth. Never blushed like that in my life,"* he said, laughing.

When asked what he cherished most about his service, Mangano replied without hesitation, ***"Being a Marine. That says it all right there. Marines are special."***

Unfortunately, Mangano lost many of his cherished photographs and memorabilia in a house fire 20 years ago. He now resides in Mashpee, Masachusetts where he continues to reflect on his extraordinary life of service.

Staff Sergeant Alfonse Mangano, thank you for your service to our great country.

Corporal Narciso "Cheeso" Massaconi

United States Army (1942–1945)

Many people tend to overuse the word amazing, but in the case Narciso "Cheeso" Massaconi its appropriate. Narciso who goes by "Cheeso" served his country admirably in the U.S. Army from 1942 to 1945. My interview with Mr. Massaconi took place in his own man cave—his garage. *"We didn't have much, but we made the most of what we had,"* the 97-year-old veteran said. *"I asked my father for a nickel to get some ice cream once...he looked at me and said, 'You wanna eat supper tonight, don't you?'*

True story."

He had two brothers who also served in the Army: Mike, who fought at the Battle of the Bulge, and Pete.

Drafted right after Christmas in 1942, the then-18-year-old Massaconi was sent to Fort Eustis in Richmond, Virginia, before deployment to the Pacific Theater. One of his first assignments was at Pearl Harbor, stationed there during its bustling and vibrant days before the infamous Japanese attack.

From Pearl Harbor, Corporal Massaconi's service took him to some of the most harrowing battles of the Pacific, including Saipan and Iwo Jima. When asked about Iwo Jima, he didn't mince words. *"It was hell on Earth. The Japanese had big guns in the tunnels on Mount Suribachi that came out on tracks. They unloaded on*

62

us…we were sittin' ducks," he said. I asked if he was scared, and he didn't hesitate: *"I was no hero…too goddamn scared to even eat…we worried about staying alive. If you wanted to die, get out of your foxhole at night."*

During the battle, Corporal Massaconi witnessed the devastating consequences of misjudgment. *"There was a black unit with us. They put those guys in tents instead of foxholes. We warned them, but the Japanese staged a sneak attack one night, throwing grenades into the tents…one by one. Killed everyone,"* he recounted with a heavy heart.

His respect for the German shepherd dogs trained for combat was profound. *"Whoever trained our dogs did a helluva job. When the Japanese killed a dog, we buried them just like one of our soldiers…they were that important."*

The chaos of Iwo Jima, with its hidden enemy and relentless danger, left an indelible mark. *"In Europe, you knew who you were fighting. At Iwo Jima, you never knew where the enemy was. They attacked when you were most vulnerable."*

The Battle of Saipan brought out equally haunting memories. *"I was on radar patrol tracking enemy planes. Our bombers would head out for missions over Japan, flying low and heavy with bombs. One time, a plane came back damaged from a mission. It dropped into the water and exploded—oil and gas burned for hours. We lost 12 young guys."*

Despite the horrors, Massaconi's sense of duty never wavered. *"I loved my country and wanted to do what was right. Still can see my mother crying, saying, 'You no go…stay home and help the family.' But I had to go."*

Now approaching his 98th birthday, Corporal Massaconi still works in his gardens and enjoys hunting deer, rabbit, and pheasant.

He's also a distinguished member of the National UNICO Baseball and Softball Hall of Fame.

Cheeso Massaconi, thank you for your service to our country.

1st Lieutenant Louise McNeill-Davis

United States Army Nurse Corps, (1939 – 1942)

1st Lieutenant Louise McNeill-Davis served as an Army Nurse in World War II from 1939 to 1942, primarily in the European Theater. Her story offers a glimpse into a crucial aspect of the war that is often overshadowed by tales of combat—the role of women who served behind the lines, providing essential medical care to wounded soldiers. The service of nurses like McNeill-Davis, especially those following General George Patton's rapid and relentless campaigns, was both groundbreaking and grueling.

Born in Beaverdale, Pennsylvania, Louise was the second of thirteen children. Growing up in a household with no running water and limited means, she found inspiration from her father, a former WWI sailor and a scholar who taught in a one-room schoolhouse. Motivated by a desire to make her father proud, Louise pursued nursing—a decision that would take her to the heart of one of history's most significant conflicts.

The Army Nurse Corps was vital to the war effort, and for women like Louise, joining meant extensive training and rigorous work. She recalled the demanding nature of their preparation: *"We had classes, then went straight to the ward to practice what we learned. In the General Hospital, you performed every phase of nursing—surgery, pediatrics, medical. We had lots of personal care responsibilities and worked in shifts covering 40-50 patients. The responsibility was immense."*

Louise's General Hospital unit followed closely behind General Patton's advancing Third Army. This proximity to the front lines meant constant movement and the ever-present threat of bombings.

Her description of their experiences likened it to the television series *M*A*S*H*: ***"We had some clowns just like the show."*** Despite the moments of levity, the reality was harsh, particularly given the limited medical supplies of the time. ***"This was before penicillin. We had lots of infections and had to give warm soaks. We used sulfur, but nothing like today's antibiotics,"*** she explained.

Working under Patton's command brought mixed feelings for McNeill-Davis. When asked about the general, she was candid: ***"I didn't like him. His attitude toward the enlisted man was very poor. He wanted those boys to be gung-ho for every event, but some of them just weren't cut out for it."***

One memory that stood out for her was the time she was tasked with caring for injured German prisoners of war in a makeshift ward set up in a schoolhouse. The emotional toll was significant, as it coincided with learning that her brother had been wounded in Anzio. Overwhelmed, she was granted a furlough to Brussels, where she found solace in the services provided by the Red Cross. ***"The Red Cross was so good. They would find sleeping quarters for you, locate restaurants for you to eat at, and even store your gas mask in case of an attack,"*** she recalled.

Women serving as nurses in WWII played a pivotal role in the war effort, and those in General Patton's units faced particularly challenging conditions. Patton's rapid advances often left medical personnel scrambling to set up hospitals in dangerous and temporary locations. The nurses were not only caretakers but also symbols of resilience and hope for the wounded soldiers. Their presence offered comfort in the midst of chaos and reminded the troops that they were not alone.

McNeill-Davis emphasized the camaraderie among the nurses and staff, despite the grim surroundings. Entertainment from the USO shows and weekly movies provided much-needed morale boosts.

But there was always a shared understanding of the sorrow surrounding them. She poignantly noted, *"We had patients that had to be back on the lines within 60 days. Our boys had lots of optimism, but those poor boys dreaded getting better because they knew where they were going."*

Despite the challenges, McNeill-Davis spoke of her service with pride. *"I was proud I did my job, and I loved it,"* she said. Her resilience and compassion were evident, especially when she spoke of the holiday season during wartime: *"You were with the same people who had the same sadness, and you couldn't let it get to you. We were too busy to think about it."*

1st Lieutenant Louise McNeill-Davis, thank you for your service to our country.

Radioman 1st Class Bob Mercurio

United States Navy (1942–1946)

Bob Mercurio served his country with honor and dedication during World War II as a Radioman 1st Class in the United States Navy. His service extended from 1942 to 1946, with an additional four years in the reserves before his discharge in 1950. At 99 years old, Bob's sharp memory and vibrant personality bring his incredible experiences to life.

Born and raised in Everett, Massachusetts, Bob was a high school football player on teams known for their strength. *"I was only 126 pounds, playing with some pretty big guys,"* he joked. Before finishing high school, he enlisted and reported for basic training at Great Lakes, Illinois, during one of the coldest winters he could remember. *"We slept with the windows cracked open, and I'd wake up with snow on my blanket,"* he recalled with a chuckle.

Bob's first assignment was aboard the USS Laramie (AO-16), a Kaweah-class fleet replenishment oiler. His first hours at sea were harrowing. *"Our ship got underway at 7 p.m. Two hours later, we were dodging torpedoes from a German submarine—it was constant,"* he recounted. Initially assigned to a 20mm gun he had no experience operating, Bob sought out his commanding officer and was reassigned.

Bob later attended Radioman School in Bedford, Pennsylvania, where he mastered the skills that would define his naval career. He was then assigned to the USS Athene (AKA-22), an Artemis-class attack cargo ship that earned two Battle Stars. His most unforgettable memory aboard the Athene came during the invasion of Iwo Jima. *"We picked up Marines on Christmas Day, did several days of maneuvers, and then anchored off Mount Suribachi. I was looking through my long glass and saw the Marines raise the flag on Suribachi. It was something I'll never forget,"* he said.

Bob also witnessed a moment of historical significance as part of the Surrender Fleet in Tokyo Bay. *"The lineup of Allied ships was the most impressive thing I've ever seen—American, British, Australian, French, even rows of battleships,"* he said with pride. He vividly remembered the end of the war while stationed in the Philippines. *"Every ship was firing its guns into the sky. It was such a great feeling."*

Among the many unique experiences Bob shared, one stands out for its lightheartedness. *"After the war, we made several trips bringing soldiers back home. On one trip, we had nine nurses aboard, including one who was very pregnant and due any day. I had 20 weeks of radioman training and sent exactly one message: letting San Francisco know we were coming in fast with a nurse about to deliver,"* he recalled, laughing.

Reflecting on his service, Bob shared a sentiment common among his generation. *"Every friend I had in high school went into the service. Your country was at war, and you wanted to do your part."*

Radioman 1st Class Bob Mercurio, thank you for your extraordinary service to our great country. Your story and sacrifices are deeply appreciated and will always be remembered.

2nd Lieutenant Regina Moskowitz

United States Army Nurse Corps (1944 – 1946)

Regina Moskowitz served her country as a 2nd Lieutenant in the United States Army Nurse Corps during World War II, a time when women were breaking barriers and redefining their roles in society. Born in Brooklyn, New York, she is a remarkable figure at 102 years of age. Though she now uses a wheelchair, her spirit remains lively, and she enjoys sharing stories of her extraordinary experiences. The youngest of five siblings, she grew up in a family steeped in service, with four older brothers who also answered the call to duty—one in the Navy, one in the Army, and two in the Marines.

Her upbringing instilled a deep sense of responsibility and compassion. She fondly recalled her parents, Itzhak and Rina Moskowitz, who created a nurturing home in Brooklyn. *"My mom and dad were two of the sweetest people you'd ever meet,"* she said. Her father managed the family clothing business, while her mother devoted herself to caring for their household. Regina shared a heartwarming story about her parents helping a young musician named Louis Armstrong. *"He would visit and play jazz at our home. They helped him get gigs whenever they could."*

As World War II engulfed the globe, Regina, like many women of her generation, felt a powerful urge to contribute. *"A bunch of us got together and decided we wanted to do our part,"* she said. At the time, the Army Nurse Corps was one of the few ways women could serve on the front lines. The Corps was an integral part of the U.S. military, but the path was not without its challenges. In 1945, the Corps faced unprecedented demands due to the scale of the war, with more than 59,000 Army nurses serving at its peak. These women worked tirelessly in combat zones, field hospitals, and evacuation stations, often under grueling conditions.

After eight weeks of rigorous training, Regina boarded a ship headed to the Pacific Theater. Her journey began with a harrowing storm. *"It was absolutely horrible. Everyone got seasick,"* she recalled. But the shared experience forged a bond among the nurses, preparing them for the trials ahead.

Regina was stationed at several Army hospitals in the Pacific, but Okinawa left the deepest impression. The battle for Okinawa was one of the bloodiest of the war, and the Army nurses were at the forefront, treating soldiers under unrelenting pressure. *"The carnage was horrific,"* Regina remembered. *"We worked 12 to 14 hours a day, seven days a week. Sleep wasn't an option. We were always covered in blood."* Women in the Army Nurse Corps were expected to perform with the same resilience and courage as their male counterparts, often facing danger themselves as hospitals came under attack.

She recalled a particularly poignant moment when a mortally wounded soldier's friend looked to her and asked, *"Ma'am, you're not going to let him die, are you?"* Regina stood by and comforted the soldier to the end. *"I wanted their parents to know that their son didn't die alone,"* she said.

Beyond the physical and emotional toll, Army nurses had to navigate their own challenges as women in a male-dominated environment. While their contributions were invaluable, they often faced skepticism and had to prove their worth in combat zones.

Even holidays offered little reprieve. *"They were tough,"* Regina admitted. *"We worked, but we thought of our families, especially at night. We had a mix of faiths among the nurses, so we celebrated everything together. That sense of camaraderie kept us going."*

When the war ended, the relief was palpable but tempered by lingering danger. *"Nobody told some of the Japanese soldiers in the hills of Okinawa that it was over, so they kept shooting at us,"* she explained. She only felt safe when her troop ship departed. *"I jumped for joy and hugged everyone in sight,"* she said with a laugh.

Reflecting on her life, Regina is most thankful for her family and the chance to have been a comforting presence for so many soldiers in their final moments. *"Knowing they saw my face instead of a ceiling or wall means everything to me."*

Regina Moskowitz and her fellow nurses were trailblazers who demonstrated unparalleled courage, resilience, and compassion in the face of unimaginable challenges, and their service saved countless lives.

Regina Moskowitz, thank you for your extraordinary service to our nation.

Private Don Palmer

United States Army (1942-1945)

Private Don Palmer served his country with valor as a combat engineer in the United States Army during World War II. Born in Hershey, Pennsylvania, in 1923, he was the youngest of three children in a close-knit family. His father, Sam Palmer, ran a grocery store in town, known for his generosity and compassion—a legacy that Don carried with him into his military service.

Don Palmer's journey in the Army began with basic training at Camp Swift, Texas, where he trained as an amphibious combat engineer. His skills in handling explosives, clearing mines, and building pontoon bridges would be critical in the dangerous missions ahead. After completing training, he was sent to England and assigned to a special unit focused on preparing for the Allied invasion of Nazi-occupied Europe—a mission that would become one of the most pivotal moments of the war.

As the days of preparation turned into weeks, Palmer and his fellow soldiers conducted endless training exercises. *"We did a ton of runs with the LCI's (Landing Craft Infantry),"* he recalled. *"We'd make trips back and forth along the English coast, practicing for the invasion."* Tension hung heavy over the soldiers as they awaited orders, knowing that the invasion would bring intense combat.

Finally, on June 5, 1944, the orders came—Operation Overlord, the largest amphibious invasion in history, was set to begin. Palmer's unit boarded their landing craft, ready to cross the English Channel. *"The morning of June 5th, we were supposed to go,"* Palmer remembered. *"But the weather was terrible—too foggy and overcast. Our Air Force planes couldn't fly, and the Navy had to postpone."*

73

The delay did little to calm their nerves. By dawn on June 6th—D-Day—the invasion began in earnest. Private Palmer's unit made their way across the channel toward the beaches of Normandy. *"I thank the good Lord every day that I wasn't in the first few waves,"* he said, visibly emotional. *"Jesus, they got massacred. The guys in those first waves... they didn't stand a chance."*

Palmer's landing craft approached the shore amidst the chaos of gunfire and explosions. *"We lost enough guys on our approach, and we weren't even in the first waves,"* he said. *"Guys were puking, getting shot... the screaming was just awful. You could feel the bullets from the German machine guns whizzing by your head."*

As his landing craft neared the beach, about 200 yards from shore, disaster struck. *"We hit a mine in the water,"* he recounted. *"The boat started to sink, and I thought to myself, 'You're never going to get out of this alive.' I had 80 pounds of equipment on me, and I thought I'd drown. I could barely move. Then, out of nowhere, a boat came by and brought us ashore."*

Once on land, Palmer faced the harsh realities of combat head-on. The Normandy beaches were littered with bodies, and the air was thick with smoke and the smell of gunpowder. *"I've never seen anything like it,"* he said. *"It was chaos."*

After surviving the initial landing, Palmer continued his mission as his unit moved through France. The combat engineers were crucial to the war effort, often working under fire to clear paths through minefields and build makeshift bridges to keep the advancing forces moving. Palmer put his life on the line numerous times, knowing that the safety of his comrades depended on his team's work.

His memories of the war are a mix of horror and beauty. He vividly recalled a somber Christmas Eve in a small French town.

"We had just lost one of our guys to a sniper earlier that day," he said, his voice trembling. *"We were all tired and sad. As we walked through the town, we saw a few lights on along the street. Most of the buildings were bombed out. We sat down to rest when I heard music—violin music—from across the street."*

Curious, Palmer and a few other soldiers followed the sound. *"We walked upstairs with our guns drawn, and the door was partially open. Inside, a man was playing 'Silent Night' for his wife and two young children—a boy and a girl. They had a single candle on the table lighting the room. They looked startled at first, but they could tell we meant no harm. Nobody spoke English, but we communicated through hand gestures. They gave us bread and cheese, and we gave the kids chocolate. The kids smiled, and the father started to play 'Silent Night' again. It was the most beautiful sound I've ever heard. I never thought that song could be that beautiful."*

Reflecting on his service and the war, Palmer spoke with a mix of gratitude and sorrow. *"I think about the guys who didn't make it,"* he said. *"We lost a lot of good men. But we did what we had to do. I'm proud to have served."*

At 100 years old, Private Don Palmer still carries the memories of those fateful days. His courage and resilience are a testament to the spirit of the soldiers who fought to liberate Europe during World War II. His experiences on D-Day and beyond are a powerful reminder of the sacrifices made by the Greatest Generation.

Private Don Palmer, thank you for your service to our great country.

Corporal Burt Paxton

United States Army (1942-1945)

Corporal Burt Paxton's service in the United States Army during World War II took him from the quiet town of Somerset, Kentucky, to the heart of Europe's battlefields. Enlisting shortly after the attack on Pearl Harbor, Paxton joined the ranks of the 6th Armored Division, a key force in liberating France and confronting the horrors of Nazi Germany. His journey offers a poignant glimpse into the courage, camaraderie, and resilience required of soldiers during one of history's darkest chapters.

After completing basic training at Fort Sampson in New York and advanced training at Fort Ord in California, Paxton was shipped to France following the D-Day invasion. His division was thrust into intense combat, including a pivotal battle in the small town of Vire. It was there that Paxton experienced a memorable encounter with General George S. Patton. *"He came in at midnight. Never seen anyone look more regal,"* Paxton recalled. *"He spoke to our unit and told us how proud he was of us. It really motivated us. I remember him wanting to see our wounded—that was really important to him."*

As a member of the 6th Armored Division, Paxton was part of the Allied push through Europe, encountering fierce German resistance. Yet, nothing could prepare him for what he would see at the Nazi death camp of Buchenwald. In April 1945, Paxton's unit was ordered to report on the camp, which had been liberated a week earlier by American forces near Weimar, Germany.

"We were greeted by 10,000-15,000 men," he said, his voice heavy with emotion. *"The only word I can use to describe the scene was horrifying. How thin they were—their bones were*

almost coming through their skin. When they smiled, it was like a smile of death."

The liberated prisoners at Buchenwald were emaciated, their bodies bearing the scars of starvation and brutality. For Corporal Paxton, witnessing the aftermath of the Holocaust left an indelible mark, shaping his perspective on the fragility and resilience of life.

Despite the horrors of war, Paxton found solace in the camaraderie of his fellow soldiers. *"We were all like brothers," he said. "Hung around together, watched each other's backs."* He shared a memory of one particularly poignant Christmas. *"A French family let us stay in their barn. They brought us food and sang Christmas songs with their two little girls. Never wished for home as much as I did that Christmas Eve."*

Now at 97 years old, Corporal Paxton reflects on his life with gratitude. *"Got my kids, grandkids—I'm content. Can't ask for much more than that,"* he said with a smile.

Corporal Burt Paxton's service in the 6th Armored Division was marked by both the triumph of liberating Europe and the heartbreak of witnessing humanity's capacity for cruelty. His story reminds us of the sacrifices made by those who served to bring freedom and justice to millions.

Corporal Burt Paxton, thank you for your service to our great country.

Corporal Jimmy Rapp

US Army (1942-1946)

Corporal Jimmy Rapp served his country in the United States Army from 1942 to 1946. At age 97, he vividly recalls the experiences and tragedies he endured during his service. Born in Topeka, Kansas, he shared his remarkable journey, filled with memories of war and the bonds formed during one of the most turbulent times in history.

Jimmy Rapp was the youngest of three siblings. His two older brothers served in the Navy and the Army, respectively, both in the European Theater. In contrast, Jimmy was deployed to the Pacific. *"I got to Pearl Harbor about three weeks after the Japanese bombed it,"* he recalled solemnly. *"It was simply devastating. The destruction—you could see the smoke from miles away. I'll never forget our ship coming into that harbor. They had welders on top of the ship trying to cut holes in it. Everyone was on the deck of our ship—nobody said a word. We were just paralyzed."*

When asked about what it meant to serve his country, Corporal Rapp became emotional. *"Outside of my wife, kids, and family, it was the greatest honor I've ever been afforded. To fight, and I mean literally fight for your country—to be able to do that, well, let me tell ya, that was something,"* he said, his voice quivering with pride.

Corporal Rapp was later sent to the South Pacific Theater, where he saw major action at the Battle of Okinawa. During this deployment, he was assigned a guard dog named Max, a loyal German Shepherd who played a crucial role in his survival.

"People have no idea how important the role dogs played in World War II. Let me tell you, that dog saved my life twice," he

78

said, his voice filled with emotion. *"Once, when I almost walked on a Japanese booby trap, and another time when I nearly got jumped by a Jap on patrol."*

Rapp shared one of the most harrowing moments of his service:

"I was on guard duty one night after two days of torrential rain. Late at night, Max started barking like crazy. I let him loose, and he took off. He barked for about twenty seconds, then I heard him yelp and whimper. The Japs had bayoneted him right through the stomach. They ran away like cowards. There was no way to help him—he was hurt too bad. I got on my knees and held him. He licked my face before he took his last breath. Nobody knows what our dogs did," he said, tears streaming down his face.

Unfortunately, the dangers faced by military dogs were immense. Like their human counterparts, they risked injury and death on the battlefield. The bond between soldiers and their canine companions was profound, and the loss of these dogs left deep emotional scars on those they served alongside. Despite their significant contributions, the sacrifices of these animals have often gone unrecognized in the broader narrative of WWII history.

When asked about life during the war, Corporal Rapp shared memories of the holidays. *"The only goddamn thing I wanted for Christmas was a shower,"* he said with a wry smile. *"I think we went three months without one. We all smelled the same. We were in the middle of everything, so you could forget about a Christmas meal. The Navy got some good stuff, my brother said—turkey, I think. We had K-rations."*

He spoke of his admiration for his older brothers, whom he considered his mentors. *"Pretty damn good guys,"* he said. *"Always there when you needed them. My brother Joe got a Silver Star. Very proud of him."*

Corporal Jimmy Rapp, thank you for your service to our great country.

Corporal Charlie Rickes

United States Marine Corps (1943-1946)

Corporal Charlie Rickes' story is one of resilience, sacrifice, and honor—a testament to the bravery of the men who served in the Pacific Theater during World War II. Born in Chariton, Iowa, Rickes grew up working on his family's farm. His strong work ethic and desire to serve his country led him to enlist in the United States Marine Corps in 1943. With two brothers already serving in the Navy and Army, he chose the Marines to distinguish himself.

After basic training at Parris Island, South Carolina, Rickes was sent to the Pacific Theater, where he would eventually find himself on one of the most infamous battlefields of World War II—Iwo Jima.

On February 19, 1945, Rickes landed on the volcanic island of Iwo Jima, a key strategic point for the U.S. military's advance toward mainland Japan. The battle would become one of the bloodiest and most brutal in Marine Corps history. Over 70,000 Marines stormed the island, facing fierce resistance from the Japanese forces entrenched in tunnels and fortified positions.

"A week later, I celebrated my 18th birthday," Rickes recalled with a bitter laugh. *"Some goddamn birthday, huh?"* The conditions on Iwo Jima were hellish. *"There was no such thing as a safe area on the whole goddamn island. Had no water. If you had to go to the bathroom, you went right in your foxhole— that's no BS."*

The landing itself was treacherous. Marines had to navigate the island's steep, black volcanic sand dunes under constant enemy fire. *"When we landed, we had to climb up these giant black sand dunes. Japs were firing at us left and right. I jumped on my belly*

81

and crawled about 25 feet next to a soldier from my hometown— shot right through the forehead. Goddamn war sucks," he said, slamming his fist on the table.

The brutal fighting left deep scars, both physical and emotional. Rickes shared the harrowing memory of a sergeant who stepped on a landmine. *"It blew one of his legs completely off, part of his other one, and part of his groin. Can still see him trying to crawl on his stomach to avoid the oncoming fire. Lost a ton of blood. His last words were, 'Kill the SOBs.'"*

When asked if he ever considered returning to Iwo Jima, Rickes' response was immediate and forceful. *"NEVER,"* he said, his voice sharp with emotion. *"Lost too many goddamn men. Thinking about it during the night just takes the living life out of you. The flashbacks are the worst—horrible memories. I can still hear the mortar fire and screams, for chrissakes."*

After the atomic bombs were dropped on Hiroshima and Nagasaki in August 1945, the war finally came to an end. However, Corporal Rickes' service wasn't over. His outfit was sent to China as part of the U.S. occupation forces tasked with stabilizing the region and overseeing the disarmament of Japanese forces.

Though his time in China wasn't marked by the same intense combat as Iwo Jima, the experience still left an impression on him. His eardrums had been punctured during the battle on Iwo Jima, but he refused to seek compensation. *"Out of respect for the fellow Marines that didn't make it home,"* he said quietly.

At 96 years old, Corporal Rickes still lives independently in Mashpee, Massachusetts, though he admits that his grandchildren check in on him regularly. *"Somebody's always sleeping over,"* he said with a laugh, grateful for the love and support of his family.

Corporal Charlie Rickes' experiences at Iwo Jima and his deployment to China are reminders of the extraordinary sacrifices made by the Greatest Generation. His service reflects the courage, endurance, and loyalty that define the Marine Corps ethos.

Corporal Charlie Rickes, thank you for your service to our great country.

Electrician's Mate Dave Rugg

United States Navy (1943-1947)

Dave Rugg proudly served his country in the United States Navy during World War II, joining the ranks of sailors who manned one of the most iconic naval vessels of the war: the Fletcher-class destroyer. Raised in Belmont, Massachusetts, he enlisted in the Navy at just 17 years old and was sent to Purdue for Officer Candidate School (OCS) training before being assigned to the newly commissioned destroyers in Los Angeles, California.

The Fletcher-class destroyers were considered the most successful of their kind, weighing 2,100 tons and designed to be fast, maneuverable, and capable of enduring significant damage. Dave Rugg served on two of these ships, the USS Dortch and the USS Cotton, both of which saw extensive action in the Pacific theater.

At just 18 years old, Dave Rugg experienced his first combat encounter—a moment that would forever be etched in his memory. *"There was a Japanese plane coming in—a Kamikaze,"* he recalled. *"He was running out of fuel. Instead of ditching in the water, he flew right into our bridge. It killed a lot of men on deck. I was below, in the engine room."*

The violent impact of the Kamikaze strike rocked the ship. *"Everything shook. The whole ship bounced around. The first thing we thought was, 'Where did we get hit? Do we have to abandon ship?'"* he remembered. The fear of the ship's seams splitting open and water rushing in was ever-present. Yet despite the danger, Rugg emphasized that there was no time for fear. *"Didn't have time to be afraid,"* he said with conviction. *"We were on the go morning and night, chasing submarines at night because they had to surface to recharge their batteries."*

Life in the engine room of a Fletcher-class destroyer was grueling. The cramped, hot space below deck was where Rugg spent most of his time. *"The only time I wasn't there was when I was sleeping,"* he explained. As an electrician, his primary responsibility was to keep the ship's generator running. *"They called me 'Sparky,'"* he added with a smile. His work was vital to keeping the ship operational, especially in high-stress combat situations.

The dangers on deck were constant, even during moments of rest. *"You couldn't smoke on deck—a lit cigarette could be seen for miles and give you away to the enemy,"* he explained. Yet there were lighter moments at sea. *"When the ship slowed down, you could go on deck and sleep,"* Rugg recalled. *"But the flying fish would jump out of the water and land on deck."* He chuckled at the memory.

One of the most historic moments of Rugg's service came when his ship was part of the protective escort surrounding the USS Missouri during the Japanese surrender ceremony on September 2, 1945. Watching from his ship as the war officially came to an end, Rugg reflected on the moment. *"It didn't really sink in,"* he admitted. After years of relentless combat, the reality of peace felt distant.

Looking back on his service, Dave Rugg described it as a whirlwind experience. *"It was fast-moving—a learning process. They told you what to do, and you did it. I was happy I went into the Navy,"* he shared. Reflecting on his fellow servicemen, he expressed gratitude for his naval service. *"I always felt the Army guys had it tough."*

Even now, at 94 years old, Dave Rugg remains a vibrant and generous individual, using his warmth and kindness to help others. His service on the Fletcher-class destroyers during World War II exemplifies the bravery and resilience of the sailors who navigated some of the most dangerous waters in history.

Mr. Dave Rugg, thank you for your service to our great country.

Private Jim Semfy

U.S. Marine Corps (1942-1946)

Jim Semfy served his country in the United States Marine Corps from 1942 to 1946 in the South Pacific Theater. He is, as one would say, a throwback to the good old days, an upstanding member of, sadly, the dwindling Greatest Generation.

We talked a bit of baseball and eased into the conversation. At 97 years of age, Mr. Semfy is extremely alert, highly intelligent, and, once you hit that vein, quite talkative.

"I could have signed a pro baseball contract in high school with the New York Giants... ah, but nobody signed back then. Everybody went in (the war)," he said. Growing up in Worcester, Mr. Semfy came from a close-knit family (he is the lone survivor of five other siblings). *"Grew up in Worcester... we were dirt poor, but we always had grub (food) on the table and plenty of bread. My mother was a great cook and could have made a shoe leather stew taste good,"* he remembered.

Mr. Semfy was drafted in 1942 and sent to Montford Point at Camp Lejeune, NC, for basic training. From there, Private Semfy was sent to the South Pacific and saw major action at the Battle of Saipan in June of 1944 as part of the 4th Marine Division.

"I'll never forget the night before the invasion... just about every guy was in the ship's chapel... next morning they gave us a big breakfast... never forget it... steak and eggs... I think everyone was in the bathroom throwing it up a half hour after they ate it," he recalled.

"We had a ton of casualties," he said sadly. *"Damn Japanese had guns mounted on the hill, snipers... the waves made it tough to*

87

hold your footing as you went from the landing craft onto the beach… couldn't keep your goddamn balance… two of my buddies got picked off just like that (snapping his fingers) as soon as we took three steps on the beach… I was pretty damn lucky… lucky."

"It was unbearable at first… the smell of death around you… guys screaming for medics that had been seriously wounded… I started screaming at the top of my lungs, just to survive… don't know why I did it… just did. I'll tell you something, those medics deserved every kind of medal you could think of… SOBs were fearless," he said slowly, shaking his head.

Private Semfy shared a terrible memory that still keeps him awake at night. *"When we got into the jungle, the Japanese would hide in caves and in tunnels… our flamethrowers would come in and let 'em have it… smelled of burned flesh… will never get rid of that smell… Jesus, it was awful… they would use civilians as targets too and use them as human shields… I… you never get over that,"* he remembered.

Another unpleasant memory he shared was when U.S. soldiers took over the island of Saipan. *"We were trying to feed women and children and this man comes up and shoves them aside… I gave him the butt of my rifle right in his goddamn head… knocked him flat on his ass… staggered to his feet and went to the back of the line… kept my eye on the SOB for a while…"* he recalled. Even at 97, Private Semfy's stare remains powerful.

I asked him about entertainment, and all he said was, *"Nah!"*

"Holidays?" I asked.

"Hurt like hell… not going to lie to you, son… hurt like hell. One thing I did do regularly was write home to my parents… nobody ever had to force me to do that," he recalled.

88

I asked him his thoughts on serving his country. ***"I'll tell ya, I'd do it again, goddamn it… we were all Americans, and that mattered back then… now today…"*** His voice trailed off.

Mr. Semfy lives in assisted living and is checked on frequently by his three children who live close by. He has six grandchildren and a great-grandchild, whom he is extremely proud of.

"I enjoyed this, quite a bit," he told me as he walked me to the door.

"Not nearly as much as I did," I said, shaking his hand.

I added this one to my book of memories. Private Jim Semfy, thank you for your service to our great country.

Sergeant Gerry Stellman

United States Army (1943–1946)

Sergeant Gerry Stellman's story represents the determined spirit of the American Forces to bring an end to the Nazi regime during WWII. Born in Madison, Wisconsin, he was the youngest of three children in a hardworking family. His father was a mailman, and his mother, whom he lovingly described as having "magic hands," was a seamstress who worked tirelessly to mend and sew clothes for their community. In 1943, at the age of 18, Gerry enlisted in the U.S. Army and began his basic training at Fort Bliss, Texas. The sweltering heat and grueling drills under the unforgiving Texas sun were a harsh introduction to military life. *"It was hot as hell,"* he recalled, remembering the salt tablets they were given to stave off dehydration in 100-degree heat.

Sergeant Stellman served in the 45th Infantry Division, known as the "Thunderbird Division." This unit, composed largely of soldiers from the Southwest, was renowned for its tenacity and effectiveness in combat. The division saw action in some of the most critical campaigns of World War II, including the push through Europe during the final months of the war.

Stationed southeast of the Battle of the Bulge in late 1944, the division faced freezing conditions. *"The weather was brutal— about a foot of snow on the ground, and the wind cut right through you,"* Gerry remembered. The psychological toll was immense, compounded by German propaganda blasted through loudspeakers in the woods. *"There was a woman's voice telling us we'd be overrun and to give up. It really got to some of the injured guys who'd come back from the front,"* he said. Despite the challenges, the 45th Infantry Division pressed forward, helping to close critical gaps in Allied lines by New Year's.

90

One of the most harrowing moments of Gerry's service came during a fierce artillery barrage. Positioned on higher ground, his unit came under heavy fire. *"The shells were coming at us like crazy. Guys were screaming—it seemed like it would never stop,"* he recalled. As the unit advanced behind a tank, Gerry was hit by artillery shrapnel. *"My right shoulder exploded… all I remember is a horrible numbness and thinking I was going to die,"* he said. A combat medic rushed to his side, treating his shoulder with sulfur powder while reassuring him to hang in there. The medic's words were a lifeline in the chaos.

Gerry was evacuated to an aid station and underwent surgery to save his arm. *"The doctor told me a bullet hit a small artery but missed major damage. I was groggy but thanked the nurse before falling asleep,"* he recounted. Recovery was a grueling process, but a moment of joy came on January 11, 1945, when he woke to find his sister Helen, an American Red Cross nurse, at his bedside. *"I yelled out 'Hel!' Everyone thought I was saying 'Hell,'"* he said with a chuckle. *"She kissed me so hard on the forehead, I nearly had a damn concussion. It was better than any CARE package I could've received."*

The hospitals where wounded soldiers were treated during WWII were a testament to both innovation and resilience. Field hospitals were often cold, overcrowded, and understaffed, but the dedication of medical personnel saved countless lives.

For his bravery and sacrifice, Sergeant Stellman was awarded the Purple Heart. Now 98 years old, he reflects on his life with gratitude. *"I've got a lot to be thankful for—wonderful children and grandchildren,"* he said.

Sergeant Gerry Stellman, thank you for your service and for reminding us of the resilience and humanity that defined the greatest generation.

Sergeant Irving "Irv" Tallon

United States Army (1943–1946)

It took months to convince Sergeant Irving "Irv" Tallon to share his story. Initially, he bluntly declined, saying, *"HELL NO!"* when asked to sit for an interview. But after building a rapport over conversations about the Patriots and his beloved Red Sox, he eventually agreed. What emerged was a story of remarkable bravery, haunting memories, and the indelible scars left by war.

Sergeant Tallon, born and raised in Worcester, Massachusetts, enlisted in the U.S. Army at 17. He was sent to basic training at Fort Belvoir, Virginia, and from there he joined the 11th Armored Division, also known as the "Thunderbolt Division," which played a pivotal role in the European Theater.

Landing in Normandy shortly after the D-Day invasion, Tallon's division pushed inland as part of the Allied advance. Normandy in the summer of 1944 was a scene of devastation. The countryside was littered with bomb craters, shattered buildings, and remnants of fierce combat.

By early 1945, the division crossed the Rhineland into Austria, capturing the city of Linz in May. It was here that U.S. forces encountered their Soviet counterparts. *"Didn't like 'em, didn't trust 'em one bit,"* Tallon recalled. *"We knew what they had done to German women and children—raping young girls... god damn animals."* His words carried the raw bitterness of those firsthand observations. Despite tensions, Tallon credited a strong-willed captain for preventing Soviet interference. *"Our captain had lots of balls and wasn't taking any crap. He set 'em straight in short order,"* he remembered.

While these encounters with the Soviets were tense, nothing could prepare Tallon and his comrades for what they found at Nazi Concentration Camp Gusen, part of the Mauthausen complex. Liberating the camp revealed unimaginable horrors. *"The smell... human, burned flesh. I will never, never forget that smell,"* he said, his voice trembling. Bodies were stacked 10 to 15 feet high, a testament to the Nazis' systematic cruelty. *"Guys who had just been in heavy combat broke down at the sight,"* he shared.

Tallon described being assigned to bury the dead, a task that left him physically ill. *"I couldn't eat for days,"* he said, shaking his head. Inside the crematoriums, soldiers found bodies waiting to be burned. Tallon recalled accompanying a colonel who stopped dead in his tracks and whispered, *"My God in heaven."* The memory of this moment brought Tallon to tears during our interview. *"Print the god damn thing,"* he insisted when I offered to omit the detail. *"I'm 96 years old... kept this inside too long."*

The liberation of Gusen was a defining moment in the war and a testament to the humanity of American soldiers. They provided food, clothing, and medical care to the emaciated survivors, many of whom were too weak to move. The soldiers' compassion contrasted starkly with the atrocities they witnessed, reinforcing their resolve to end the Nazi regime.

Tallon spent the remainder of the war helping to stabilize Austria and ensure peace. Reflecting on his service, he admitted, *"I thought nobody cared... didn't want to listen."* But Sergeant Tallon's experiences are a vital part of history, and his courage—both on the battlefield and in finally sharing his story—deserves recognition.

Sergeant Irving Tallon, thank you for your service and for bearing witness to one of history's darkest chapters. Your bravery and sacrifice will never be forgotten.

Lieutenant Myron Walden

United States Army Air Corps (1942–1945)

It was an honor to interview Lieutenant Myron Walden who served with the United States Army Air Corps during World War II in the European Theater. Drafted at the age of 19, Myron left his home in Springfield, Massachusetts, to embark on a journey that would shape the rest of his life.

Lt. Walden's military journey began at Fort Dix, New Jersey, where, as he jokingly put it, *"they taught you how to salute."* From there, he traveled to Miami Beach, Florida, for six to eight weeks of basic training, followed by assignments in Long Beach, California, for gunnery school, Salt Lake City, Utah, for small arms training, and Las Vegas, Nevada, for airplane recognition. The transition from a civilian to an airman wasn't easy. *"The first time I flew, I got sick and vomited all over myself. I thought, 'Some pilot you are,'"* he chuckled. Despite the rough start, earning his wings was a moment of pride. *"I was thrilled. I was a Waste Gunner. I thought it was pretty neat—Clark Gable was a Waste Gunner!"* he said with a smile.

The reality of war quickly set in as Lt. Walden's squadron averaged three missions a week. *"Our first mission was bombing a cement factory in Germany—no problem. But on the third mission, we hit a factory, and on the way back, the plane next to us got hit. It was my good buddy's plane, and it disintegrated in mid-air. Just like that,"* he recalled solemnly.

One haunting memory from late 1943 has stayed with him ever since. *"I was young and hot-headed and snapped at our co-pilot. I didn't like him—always thought he was anti-Semitic. They grounded me for insubordination. While I was grounded, my squadron went on a mission and never came back. They were*

94

shot down over Germany. It haunts me to this day. I had some really good pals on that plane," he said quietly.

Lt. Walden flew a total of 44 missions during his service, but none were without risk. He vividly remembered his final mission. *"We engaged in an air fight with the Germans. Planes were everywhere—it lasted about 15 minutes. We lost nine B-17s and nearly 100 men. When we got back to base, I looked at our plane. There was a 20 mm shell lodged just above the wing— unexploded. I backed away 500 feet, dropped to my knees, and kissed the ground,"* he recalled. In total, his squadron lost 79 planes during their time in Europe.

When asked about holidays and entertainment during the war, Lt. Walden was frank. *"Holidays? We had no god damn time to celebrate or even think about them. All you thought about was preparing for the next mission,"* he said. Entertainment was equally scarce. *"Cards—that's the only entertainment we had,"* he scoffed.

Reflecting on the end of the war, Lt. Walden's response was simple but profound. *"Relieved. I flew 44 missions. I was just happy to get my ass home."*

Now 95 years old, Lt. Walden is a widower living in Bourne, Massachusetts, where he enjoys the love of his four children and seven grandchildren.

Lieutenant Myron Walden, thank you for your extraordinary service to our great country. Your story and sacrifices will never be forgotten.

Korean War Veterans Spotlights (1950–1953)

Veterrans Spotlights

Lt. Colonel Dorothy Courtemanche – United States Army Nurse Corps (1950–1973)
Private Cam Davidson – United States Army (1952–1953)
General Gordon R. Sullivan- United States Army (1959-1995)
Colonel James L. Tow – United States Army (1952–1980)

The Korean War:
Grit, Sacrifice, & Resolve

The Korean War was a brutal conflict that shaped the geopolitical landscape of the modern world. The active fighting ceased in July of 1953. However, no formal peace treaty was ever signed between North and South Korea and the DMZ, (the 160 mile stretch that separates the two country's) is still one of the most heavily fortified pieces of land on the planet.

It was a war of extremes with searing summers followed by bone-chilling winters, battles waged in dense mountains and war-ravaged cities, and an enemy that refused to break. For the men and women who served, it was a war of survival, sacrifice, and unyielding determination, one that did not receive the recognition it deserved. The three years of combat killed over 1.5 million North Korean's and almost a million South Korean's, but the real toll was represented by the death of two to three million civilians (over 36,000 American service members lost their lives in the Korean War).

Enduring the Elements

The Korean Peninsula was unforgiving, both in terrain and climate. Soldiers arriving in the summer of 1950 faced monsoons that turned battlefields into mud-soaked quagmires, while winter brought temperatures as low as -30°F.

For the medical personnel, the weather made saving lives an even greater challenge. **Liuetenant Colonel Dorothy Courtemanche**, an Army nurse, described the relentless nature of wartime medicine. *"You had to be self-sufficient,"* she explained. *"I had 20 soldiers on each side of my ward. You didn't call the lab to get blood—you drew it yourself. You got competent real fast. The* *soldiers I took care of were so upbeat. They never complained. It was such an honor to care for them."*

Kindness and Respect for the Korean People

Many American soldiers felt a deep sense of compassion for the suffering of the Korean people. With cities bombed out and entire villages destroyed, thousands of civilians were left homeless, hungry, and vulnerable to the harsh Korean winters. American military medics and nurses treated wounded civilians when possible.

Lt. Courtemanche recalled the professionalism of Korean nurses working alongside her: *"They were incredibly skilled and dedicated. They didn't socialize much with us, but they worked tirelessly."*

Private Cam Davidson's time in Korea was also enriched by moments of joy and creativity. An accomplished pianist since childhood, he discovered an old piano at the orphanage and quickly put it to use. *"I tuned it up as best I could and taught the kids to sing nursery rhymes and fun songs. We even organized plays—it brought so much happiness to those children."*

The Reality of War, Reflected in *M*A*S*H*

For many Americans, their only exposure to the Korean War came decades later, not from history books, but from their television screens. The long-running series *M*A*S*H* (1972-1983), though often humorous, captured many of the war's grim realities: the relentless influx of wounded soldiers, the exhaustion and emotional toll on medical personnel, and the camaraderie that kept them going.

The finale of *M*A*S*H*, which aired on February 28, 1983, was watched by over 105 million viewers, making it the most-watched television episode in history outside of the Super Bowl.

Their Legacy Lives On

The Korean war is often referred to as "The Forgotten War," The men and women who served in Korea did not fight for medals or recognition. They fought for their fellow soldiers, for the ideals of freedom, and for a nation on the brink of collapse. They endured the cold, the bloodshed, and the silence that followed, knowing that their sacrifices mattered even if history too often overlooked them.

This book seeks to change that. To ensure their voices are heard in their own words. To remind us that they were there, and that their sacrifices will never be forgotten.

Lieutenant Colonel Dorothy "Dot" Courtemanche

United States Army Nurse Corps, (1950-1973)

Lieutenant Colonel Dorothy "Dot" Courtemanche devoted 23 years of her life to military service as a nurse in the United States Army. Her remarkable career took her from training in Augusta, Georgia, to hospitals and military bases across the globe, including an unforgettable assignment at the 121st Evacuation Hospital in Seoul during the Korean War.

Arriving in Seoul, Korea, Lt. Colonel Courtemanche was immediately thrust into the harsh realities of wartime medicine. The 121st Evacuation Hospital was the lifeline for countless soldiers wounded on the frontlines. The summers were unbearably hot, and the winters brought freezing temperatures that made living conditions difficult. *"The monsoons were brutal,"* she recalled. *"The whole base would flood. We had six nurses to a hooch—a small, cinder block building. We wrote home asking our families to send electric blankets just to get through the winter nights."*

The conditions weren't the only challenge. The war-torn city of Seoul carried the lingering scent of septic systems damaged in the conflict, a smell that became an inescapable part of life. But the spirit of the Korean people left a lasting impression on her. *"They were lovely people—very respectful,"* she said. *"The Korean nurses were incredibly skilled and professional. They didn't socialize with us much, but they were dedicated to their work."*

101

Working in a combat hospital during the Korean War demanded resilience, adaptability, and quick thinking. *"You had to be self-sufficient,"* she explained. *"I had 20 soldiers on each side of my ward. You didn't call the lab to get blood—you drew it yourself. You got competent real fast. The soldiers I took care of were so upbeat. They never complained. It was such an honor to care for them."*

She remembered how important it was to lift their spirits, even in the darkest moments. *"I'd walk in and say, 'Hello, everybody,' and you could feel the mood in the room change,"* she shared. The camaraderie she built with her fellow nurses and patients was a powerful bond that carried them through the hardships of war.

Serving during the holidays was especially difficult. *"Christmas was tough,"* she recalled. *"Nobody wanted to be alone, so everyone signed up to work. We had a Christmas tree in the hospital, and it brought a bit of comfort to all of us."* One of her fondest memories was seeing Bob Hope perform for the troops. *"I almost missed it,"* she laughed. *"I'd just gotten off a night shift and wanted to sleep. But my friends insisted I go. The show was incredible. Bob Hope was 64 years old at the time—he sang, danced, and entertained us. It was such a morale booster, but when they ended with 'Silent Night,' everyone was crying."*

However, the hardships of war weren't always so easily softened. One memory, in particular, stood out to her: *"It was Christmas time in Korea, and we had six overdose deaths in the hospital. The hospital couldn't accommodate the bodies—they were already overloaded—so they placed them in boxes outside in the freezing cold. Knowing those six bodies were out there was heartbreaking."*

Despite the trauma, Lt. Colonel Courtemanche found solace in the close-knit community of nurses. *"The camaraderie we shared was the highlight. We saw so much death, but we were in it together.*

If it weren't for that, we all would have gone nuts. On our days off, we visited other MASH units to check in on our friends."

Working in the ICU meant dealing with a wide variety of cases. *"We handled everything—from a colonel who got drunk and fell off his barstool to a baby in an incubator,"* she said. Her transition to ICU work was guided by her mentor, Captain Eileen Gentile, whom she credits with helping her grow into a more confident nurse. *"I was terrified at first—afraid I'd hurt someone. But Captain Gentile was patient and supportive. She became like a big sister to me."*

The ICU wasn't without its lighter moments. She recounted one humorous story of a soldier desperate to get home for his anniversary. *"He wasn't supposed to move, but he went to the latrine and jumped out the window. I was a wreck telling my supervisor what happened. She had the MPs bring him back. I felt terrible, but we all had a good laugh afterwards."*

Reflecting on her career, Lt. Colonel Courtemanche expressed deep gratitude for her time in the service. *"It went by so fast. I miss it a lot. I wouldn't have traded it for the world."*

Lt. Colonel Dorothy Courtemanche's service during the Korean War and beyond exemplifies the dedication, resilience, and compassion of military nurses who cared for soldiers in the most challenging circumstances. Her legacy is one of unwavering commitment to both her country and the men and women she cared for.

Thank you, Lieutenant Colonel Courtemanche, for your service to our great country.

Private Cam Davidson

United States Army (1952–1953)

Private Cam Davidson served his country with dedication and compassion during the Korean War from 1952 to 1953. Born in Fort Worth, Texas, he enlisted in the Army Reserves in 1951 and completed basic training in Cheyenne, Wyoming. Excelling in testing and demonstrating exceptional skills, he was selected for Army Military Intelligence and subsequently deployed to Korea.

Upon his arrival in Pusan, Private Davidson was struck by the devastating economic conditions caused by the war. One of his earliest memories involved a young boy who stole his wallet. *"This little kid could've won an Academy Award,"* he recalled with a laugh. *"He kept a straight face, swearing up and down, 'I no take wallet,' even though it was bulging out of his back pocket. He actually negotiated some chocolate before giving it back."* Reflecting on this moment, he added, *"I don't know why, but I felt that encounter with the boy was the start of something meaningful."*

In addition to his military duties that included translating documents, managing communications, and even working as a telephone operator, Private Davidson found a calling beyond the battlefield. He volunteered at a local orphanage and a nearby medical clinic in Pusan, both run by a dedicated group of nuns. *"These women were incredible. They were light-years ahead in their techniques and completely selfless,"* he said. The head nun, Sister Margaret, was affectionately dubbed "The General" for her tough but compassionate leadership. *"She didn't take any nonsense from anyone,"* he added.

Private Davidson's time in Korea was also enriched by moments of joy and creativity. An accomplished pianist since childhood, he discovered an old piano at the orphanage and quickly put it to use. *"I tuned it up as best I could and taught the kids to sing nursery rhymes and fun songs. We even organized plays and it brought so much happiness to those children."* However, one project left an indelible mark on his heart.

"We put together a choir for Christmas," he said, his voice softening. *"There were four sisters, ages 9, 12, 14, and 16, who had beautiful voices. Their father had planned to sell them into prostitution, but a relative rescued them and brought them to the orphanage. They loved to sing 'Silent Night,' and we were preparing for a Christmas play."* His tone turned somber as he recounted the tragic turn of events. *"On December 20th, Sister Margaret and some nuns went to get supplies despite warnings of active fire. The enemy bombed the orphanage while they were away. Everyone there was killed—all the kids and two nuns."* Fighting back tears, he continued, *"Every time I hear 'Silent Night,' I think of those beautiful kids. It breaks my heart. They knew nothing about the war, and yet they paid the ultimate price."*

The loss continues to haunt him, even decades later. *"All these years, and it still gets to me. Those poor kids did absolutely nothing to deserve that,"* he said, shaking his head. *"I've never spoken about it—not even to my family."*

Now 88 years old, Private Davidson lives in Mashpee, Massachusetts, where he still plays the piano. His story is a poignant reminder of the human cost of war and the enduring impact of acts of kindness amid unimaginable hardship.

Private Cam Davidson, thank you for your service to our great country. Your compassion and dedication will never be forgotten.

General Gordon R. Sullivan

United States Army (1959-1995)

What an honor and true pleasure to sit down and interview General Gordon R. Sullivan. The former Chief of Staff of the United States Army and four-star General graciously welcomed me into his beautiful home. As General Sullivan brought me into his "office," I was overwhelmed by the history that surrounded me, and framed pictures, flags, awards, and commendations filled the room, each one a testament to his remarkable career. I asked about the stars on one of his flags, a Vice Chairman's flag, and he explained, *"There are thirteen of them. They represent the thirteen original colonies."* It was a powerful reminder of the enduring legacy of our nation's founding. Among the treasures in his collection were extraordinary books on military strategy and history, including his own, *Hope Is Not a Method*.

Published in 1996, *Hope Is Not a Method* reflects General Sullivan's visionary leadership during his tenure as Chief of Staff. The book emphasizes the importance of innovation, adaptability, and clear purpose in achieving success. Drawing on his experiences during a transformative period in the U.S. Army, Sullivan offers timeless lessons in leadership that resonate far beyond the military.

The most striking quality of General Sullivan is his humility. Despite his towering military stature, he is one of the most down-to-earth individuals I've ever met—a sentiment echoed by many who see him at local coffee shops. As I took my seat, the General handed me a copy of the Norwich University "Statement of Guiding Values." A proud graduate of the Class of 1959, Sullivan

106

embodies the university's core values: integrity, honor, service to the nation, and perseverance in adversity. *"These values never change,"* he told me, underlining their importance.

He shared a touching story from his time at Norwich, a university renowned for its legacy as the birthplace of the Reserve Officers' Training Corps (ROTC). Norwich, founded in 1819, is a cornerstone of military education in America and has produced numerous distinguished graduates, including Admiral George Dewey, hero of the Battle of Manila Bay. Sullivan's journey at Norwich was not without challenges. *"I worked in the kitchen to earn extra money. My mother told me she couldn't afford college anymore. One of my classmates, who signed with the Boston Red Sox, co-signed a loan for me. I worked as a summer bartender on the Cape to pay off my debt,"* he recalled.

General Sullivan's desire to be a soldier took root at ROTC summer camp at Fort Knox in 1958. *"I loved soldiering,"* he said. This passion carried him through an illustrious career spanning more than three decades. After commissioning into the United States Army as an Armor officer, Sullivan began a path that would take him around the world. He served a 14-month deployment in Korea, where he gained firsthand experience of the challenges and intricacies of operating on the Korean Peninsula.

He later completed two tours in Vietnam during one of the most tumultuous periods in American military history. His first tour saw him serving as a company commander in the 3rd Battalion, 22nd Infantry Regiment, 25th Infantry Division, where he led his troops through the harsh conditions and unpredictability of jungle warfare. During his second tour, he served as a staff officer, honing the organizational and operational skills that would define his later career. His leadership earned him the respect of his peers and subordinates alike.

Sullivan's service also extended to Europe, where he commanded an armored brigade as part of NATO's efforts during the Cold War. His career culminated in his appointment as the 32nd Chief of Staff of the United States Army in 1991, a position he held until 1995.

When I asked about leading during the holidays, Sullivan shared his deep care for his troops. *"It was a big concern for me. We tried to keep soldiers entertained and out of the fields during Thanksgiving and Christmas, giving them a little feel of home. You cannot let your soldiers feel like you don't love them. They need to know that you respect them."* He then read aloud a moving letter from General Ulysses S. Grant to General William T. Sherman, dated March 1864, which spoke to the profound bond of mutual trust and sacrifice among soldiers. *"What it means,"* Sullivan explained, *"is that you're going to die trying."*

During our time together, I also had the privilege of exploring *Gordon R. Sullivan, The Collected Works 1991–1995*, a compilation of his notes, speeches, and personal reflections. It became clear that General Sullivan was never a "desk" general. His love and passion were with his troops, and he remains a steadfast advocate for their well-being.

General Gordon R. Sullivan's remarkable career, his contributions to military thought, and his dedication to his troops and the nation are truly inspiring. Norwich University, the Army, and the United States have been profoundly shaped by his leadership and service.

General Sullivan, American hero and patriot, thank you for your service to our great country.

Vietnam War (1955–1975)

Veterans Spotlights

Captain Bill Anners- United States Army (1968–1970)
Scout Pilot John Aoila- United States Army (1967-1971)
Sergeant Richard Batista- United States Army (1968-1970)
Seaman Curtis Beal- United States Navy (1969-1971)
Specialist 4 Combat Medic Bill Blaisdell- United States Army (1969–1970)
Corporal John "Jack" Demello Jr.-United States Army (1967–1969)
Specialist 4 Dave DiMestico- United States Army, (1970-1971)
Sergeant Al DiMuzio- United States Army (1967-1970*)*
Colonel Jack Donovan- United States Marine Corps (1964-1966)
Specialist E-4 Bob Dutra- United States Army (1969)
Captain John Ferguson Sr.-United States Air Force (1958–1969)
Airman 1st Class Tim Flynn – United States Air Force (1965–1969)
Sergeant Barry Funfar – United States Marine Corps (1966–1970)
Lt. Colonel John Harris-United States Air Force (1966-2000)
Captain Carter Hunt – United States Marine Corps (1966–1969)
Dr. Jim Kenner-United States Army (1967-1970)
Sergeant Dick McCarthy-United States Army (1966-1968)
Captain Mark McMahon - United States Marine Corps (1968–1973)
Sergeant Major John Mennitto – United States Army (1965–1989)
Etta Moran-American Red Cross Donut Dollie (1967–1970)
Sergeant Carl Rosetti- United States Army (1967–1969)
Specialist 5 Dave Smachetti- United States Army (1965-1967*)*
1st Lieutenant Brian Sullivan – United States Army (1970–1971)
Lt. John Shoemaker – United States Army (1968–1972)
Corporal Walt Stillman – United States Army (1967–1970)

110

Staff Sergeant Kris Tebbetts – United States Air Force (1965–1968)

Combat Medi Billy Ray Thomas- United States Army (1968-1969)

Colonel James L. Tow – United States Army (1952–1980)

Sergeant Don Silvia – United States Army (1965–1968)

Colonel Bob Walsh – United States Air Force (1968–1998)

Seaman Dave Westcott- United States Navy (1964-1966)

Staff Sergeant Arthur Wiknik – United States Army, 101st Airborne (1968–1970)

Haunted Heroes:
Forgotten and Forsaken

For our military veterans who fought in Vietnam, the war did not end when they left the jungle. It followed them home in nightmares, in memories, in the silence of a country that often refused to acknowledge their sacrifice. Unlike the wars that came before, there was no ticker-tape parade, no national celebration of victory. Instead, they returned to protests, indifference, and a lingering question: *Was it all for nothing?*

Yet, through their words, we find a truth that history sometimes overlooks. These men and women did not fight for politics, they fought for the person beside them. They fought because they were sent. And for those who survived, they carried the war in their memories long after the last helicopter lifted from Saigon.

"The Living Room War"

Television had a profound impact on the public's perception of the Vietnam War and it was the first major conflict broadcast directly into American homes. Unlike previous wars where news was delayed and/or often sanitized, television brought raw real-time footage of combat, bombings, wounded soldiers, and perhaps more importantly wounded civilians into people's living room.

Journalist like CBS's Walter Cronkite, once considered the most trusted man in America, publicly question the war, and on one occasion, he told his audience of millions that the war could not be won. His view departed from the government official optimism and influenced public opinion. President Lyndon B. Johnson reportedly said, If I've lost Cronkite, I've lost Middle America.

A War Without Front Lines

Vietnam was not a conventional war. There were no clear front lines, no safe zones. The enemy did not wear uniforms, and the jungle itself seemed alive with danger. **Lieutenant John Shoemaker** described his first patrol as a baptism by fire, facing an ambush before he even had time to understand where the shots were coming from. *"We were marching through a rice paddy; I was fifth in line and the radio operator was in front of me. I was mesmerized as I waded through dark brown, putrid, god-awful-smelling water filled with water buffalo and human excrement, and leeches."* A soldier ahead of him stepped on a booby trap, causing a massive explosion. *"The blast severed one man's foot and nearly split the lieutenant in two, "Suddenly, I was in charge."*

For many, the enemy was not just the Viet Cong but the very land itself. The dense jungle, the monsoon rains, and the suffocating heat made survival a daily battle. **Sergeant Don Silvia** recalled how booby traps and sniper fire turned every step into a potential death sentence. *"We were in Jump Battery, supporting the infantry. It was monsoon season, and we had to abandon our guns and position in the field. The VC knew there would be flooding. We followed the ARVN (Army Republic of Vietnam) up the mountains. It rained all night long. We had tents, but nobody slept. The next day we're back and all our guns were sunk in the mud."*

The Helicopter War

Above the jungle, in the skies, helicopter pilots and door gunners played a vital role in extracting troops, providing fire support, and saving the wounded. The Vietnam war marked the first large scale use of helicopters in combat. The Bell UH-1 Iroquois (or the "Huey" as it was known) was used extensively for air cavalry operations, allowing soldiers to be dropped into remote jungles.

113

During the war nearly 7000 Huey's were used in Vietnam and over 3000 were lost in combat.

Sergeant Barry Funfar was a Marine door gunner on a Huey. His missions included night flying, medevacs, extractions, and providing critical air support to ground troops. *"They loved seeing us come to get them. The life expectancy of a door gunner in combat was just five minutes, but you didn't think about it. Adrenaline takes over, and you do what you're trained to do."*

The Unwelcome Homecoming

For many Vietnam veterans, the greatest shock was not the war—it was coming home. Unlike their fathers, uncles, and friends who returned from World War II as heroes, they were often met with hostility, resentment, or worse, total indifference.

Sergeant Al Dimuzio commented, *"when I returned home, I remember walking through the airport, and someone spit at my feet. That person had no idea what I had gone through. My thought was, '(Expletive) you, pal.' Despite that, I was grateful to be home and confident I had fulfilled my duties. I chose to ignore the negativity and focus on my pride in serving."*

This hostile reception was brutally depicted in **Oliver Stone's "Born on the Fourth of July," starring Tom Cruise,** which was based on the true story of Vietnam veteran Ron Kovic. The film captured the stark reality that many veterans faced upon returning—a nation that turned its back on them. Kovic, who was paralyzed in combat, returned home expecting gratitude for his service, only to find himself ridiculed and abandoned. He was spit on, called a "baby killer," and left to fend for himself in a broken VA system that offered little help to wounded veterans.

Marine Captain Carter Hunt may have described it best when asked about the anti-war protests back home, he responded thoughtfully. *"That's what we were defending—the right to free speech, religion, and more. What bothers me is the people who went to Canada and didn't serve in any capacity. I believe in universal service—whether it's the military, Peace Corps, nursing homes, something."*

The Psychological Toll

Even after leaving the battlefield, the war lived on in the minds of those who fought it. PTSD, though not yet widely understood, haunted countless veterans. **Sergeant Carl Rosetti** still wakes up in cold sweats from nightmares that have never faded. *"I see their faces. The guys we lost. The ones who didn't make it."*

Sergeant Richard Batista witnessed his friend, **Lt. Bob Poxon**, sacrifice himself for his unit. *"It was June 2, 1969, in the Province of Tay Ninh, we got dropped in a landing zone and came under intense fire…one of our guys got hit…Lt. Poxon went to help the soldier and got hit himself…told us to concentrate our firepower on an enemy bunker he identified…gets up and runs toward the bunker and pulls the pin out of his grenade…gets shot a second time..throws the grenade into the bunker but gets shot right before which proved to be fatal."*

The struggle to adjust to civilian life was compounded by a society that didn't want to hear their stories. Many turned to alcohol, drugs, or isolation. Some, unable to cope, took their own lives. Vietnam was one of the first wars where PTSD was widely recognized, yet it was decades before many veterans received the help they needed.

The Brotherhood That Never Faded

Yet, despite the hardship, despite the politics, the one thing that remained was the brotherhood. Veterans of Vietnam didn't fight for glory, or even for country. They fought for the man beside them.

Sergeant Al DiMuzio describes the camaraderie among his fellow soldiers as a positive and unifying experience. *"It wasn't 100% consistent, but I had the opportunity to get to know people of different religions, beliefs, backgrounds, and cultures, and to embrace that diversity. The camaraderie and commitment were always there. We'd get mortared and shelled, but the bond with my fellow soldiers kept us going."*

Their Stories Will Not Be Forgotten

For too long, Vietnam veterans were cast aside, their service ignored or overshadowed by the political turmoil of the time. But their sacrifices were real. Their struggles were real. And their courage and service to our country deserves to be remembered. These are their stories in their own words.

Captain Bill Anners
United States Army (1968–1970)

Bill Anners, a Vietnam War veteran whose extraordinary service as a helicopter pilot in the 1st Air Cavalry Division remains an inspiring testament to courage under fire. Born and raised in Worcester, Massachusetts, Bill comes from a proud lineage of military service, his grandfather served in World War I, and his father fought in World War II in North Africa.

After completing helicopter flight school in the United States, Bill found himself, at just 18 years old, flying Cobra helicopters, a sophisticated, million-dollar aircraft. *"We were the forward eyes of our ground troops, providing fire support and reconnaissance. To be in the Air Cav, you needed brass balls,"* he said with a mix of pride and solemnity. *"Your door gunner and observer were everything—they had to be good men."*

Stationed at Camp Holloway in Pleiku under the command of Lieutenant Colonel John Hughes, Bill's missions ranged from critical fire support to life-and-death reconnaissance. Early in his service, he experienced the brutal realities of war. *"On my second mission, a Viet Cong sniper shot through the windshield, hitting my co-pilot in the neck. Blood was everywhere, but I managed to get him to a MASH unit in time. He survived, but he lost a lot of blood."*

The dangers of flying in the 1st Air Cavalry were ever-present, and losses were frequent. One particularly harrowing memory involved a night mission gone awry. *"Scout teams lost a lot of men"* he said quietly. Mr. Anners became silent for about 3 minutes, and I gave him his space.

117

"We got hit and had to crash-land our helicopter right in the jungle. We ran like hell for cover and spent 10 hours waiting for support. At one point, I saw a Bengal tiger no more than 100 feet away, stalking through the trees. We were scared shitless. The Viet Cong could've taken us, but they didn't. They always waited until you weren't ready to attack."

Moments of levity were rare but cherished. Bill recalled seeing Bob Hope perform at Camp Holloway in 1969. *"We'd been on constant patrol and were wiped out, but security was tight, and Hope brought Anita Bryant, Miss World, and Raquel Welch, who danced her heart out. I was in the front row and could've touched her! After the show, Bob stayed with us for two hours, just talking and joking. What he did for us—I'll never forget it."*

Coming home from Vietnam was a stark contrast to the camaraderie of his service. Bill recounted a painful memory from his return to the United States. *"When I landed at the airport in San Francisco, a woman came up to me, spit on my shoes, and cursed me. We got absolutely nothing when we came home. But I have no regrets. A soldier does what he has to do."*

Despite the lack of recognition upon his return, Bill's service earned him an impressive array of honors: 12 Air Medals, two Bronze Stars, a Purple Heart, and the Distinguished Flying Cross. Reflecting on his time in Vietnam, he said, *"I'm glad I did it. Those SOBs burning flags back home wouldn't have lasted a day in Vietnam."*

Captain Bill Anners, thank you for your extraordinary service to our great country. Your bravery and sacrifices are deeply appreciated and will never be forgotten.

Scout Pilot John Aoila

U.S. Army (1967-1971)

John Aoila served his country in the United States Army from 1967 to 1971 as a Scout Pilot, flying over 200 missions during his service. Born in Fort Kent, Maine, Mr. Aoila was drafted in the summer of 1967. *"I grew up on a potato farm... I was the oldest of four... we had potatoes morning, noon, and night,"* he said with a laugh.

Still in phenomenal shape at age 74, he attributes his fitness to walking a beat as a policeman for 38 years. *"I walked in my sleep,"* he said, laughing. However, the humor ended quickly when he shared his thoughts on serving in Vietnam. *"We were treated like (expletive) when we came home... the combat never really bothered me, it was the way I was treated when I came home after serving my country... came into San Francisco... they told us to expect some protests... a group of women spit right at me... actually SPIT at me,"* he recalled with disgust.

"Guys I served with were... foxhole guys... when you went out on a mission, you knew they had your back, every time." He shared a story that painted a vivid picture of his time in combat: *"Our mission was to provide cover to our infantry companies. We flew H-13 helicopters, a pilot and a gunner. Each helicopter had M-60s machine guns onboard. You had to know what the hell you're doing, or you were done in a matter of seconds. It was nerve-wracking work. We were always flying low, just skimming above the treetops to keep our boys safe on the ground. The enemy knew how to target us, and we were always in their sights."*

He continued, *"We had a mission and were flying in front of our troops... VC fired at us... we got hit, and I couldn't keep the*

119

helicopter in the air... hit the ground hard and crashed... we jumped out and dove into a ditch... blood was coming from my mouth... broke my jaw. We could hear the enemy coming at us... I was scared beyond (expletive). Then our boys rose up (helicopters) over a large hill and mowed 'em down. You never know how close you are to... dying... it's just an awful feeling."

Flying H-13 helicopters was no easy task. The compact, bubble-canopy aircraft offered little protection from enemy fire, making every mission a harrowing experience. Pilots had to master quick maneuvering and stay alert to ground threats. The role of providing cover was critical to the survival of ground troops, and Scout Pilots like Mr. Aoila often flew in the most dangerous conditions, knowing full well that any wrong move could mean disaster. *"We were the eyes of the infantry. You had to trust your instincts and your crew. It wasn't just about flying—it was about staying alive long enough to do your job."*

When asked how the holidays were in Vietnam, he was candid. *"Honestly? I got drunk as a skunk... all of us did... I missed home... nothing I could do about it. We played cards and drank beer... one Christmas, we sung a drunken rendition of 'White Christmas' as we staggered back from the club,"* he remembered.

When I asked him if he had any animosity towards the protestors, he said this: *"I've mellowed quite a bit... there was a day that I would have taken them out... wouldn't have bothered me one bit... they had no clue... we were doing what we were told to do... that's what a soldier does. Saw an awful lot of (expletive) in 'Nam... I didn't ask to see guys blown up in front of me or to see my buddies get killed... I blame the government... they hung us out to dry. If the politicians that start the wars had to fight 'em, we wouldn't have any goddamn wars,"* he said.

Mr. Aoila, a highly decorated combat veteran, lives with his wife Esther in Mashpee, Massachusetts. They have three daughters and four grandchildren.

Mr. John Aoila, thank you for your service to our great country, and welcome home.

Sergeant Richard Batista

U.S. Army (1968-1970)

Sergeant Richard Batista served his country with unwavering dedication during the Vietnam War from 1968 to 1970. His journey, marked by courage and leadership, highlights the resilience of those who endured one of the most challenging conflicts in American history.

Sergeant Batista began his military career at Fort Dix, New Jersey, where he completed basic training. He then went on to Fort Polk, Louisiana, for advanced training before being deployed to Vietnam in late 1968. Assigned to the 1st Air Cavalry, he arrived in the dense jungles of A Shau Valley equipped with an M-16 rifle, a pistol, ammo clips, and essential gear. His first impression of Vietnam was stark: *"Frightening... you didn't know what to expect,"* he recalled.

After his initial deployment, Sergeant Batista was stationed in Phu Vinh, where he quickly distinguished himself as a leader and was promoted to sergeant. *"I was always a leader. I jumped two ranks,"* he said. His focus on discipline and readiness set him apart. *"Lots of guys were smoking weed, drinking a lot to numb the pain... I never did those things. I had to be alert, focused, and in a good frame of mind. If not, people got killed."*

Sergeant Batista emphasized three key qualities of effective leadership: *"Stay safe and do the right things. Make sure your guns are always cleaned to prevent jamming in a firefight. Treat everyone with respect and dignity."*

When asked about spending the holidays in Vietnam, he shared how soldiers coped with being away from home. *"You learn to deal with it... just another day. I saw Bob Hope a couple of times, which was good."*

One of the most emotional moments in Sergeant Batista's service was witnessing the heroism of his friend, Lt. Bob Poxon. He recounted a harrowing day in June 1969 in the Tay Ninh Province. *"We got dropped in a landing zone and came under intense fire. One of our guys got hit. Lt. Poxon went to help the soldier and got hit himself. He told us to concentrate our firepower on an enemy bunker he identified. Then he gets up, runs toward the bunker, and pulls the pin out of his grenade. He gets shot a second time... throws the grenade into the bunker but gets shot right before, which proved to be fatal."*

Sergeant Batista was visibly moved as he recounted the bravery of his fallen comrade. *"Lt. Poxon's courage and indomitable will are something that I will always admire and never forget."* (Lt. Bob Poxon was posthumously awarded the Medal of Honor.)

Returning home from Vietnam was another profound experience for Sergeant Batista. He vividly described the atmosphere aboard the packed TWA flight carrying soldiers back to the United States. *"As soon as we took off, there was a thunderous roar of screaming and yelling. It didn't stop until we landed in Oakland seventeen hours later."*

Back home in Boston, Sergeant Batista recalled a touching gesture from his hometown mayor. *"I was home for about a week or two, staying at my parents' house. The doorbell rings on a Saturday morning. I answer it, and it's the Mayor, James McIntyre. He hands me a fruit basket. He did that for all the veterans in town that morning. What a man he was."*

Reflecting on his service, Sergeant Batista expressed immense pride in his contributions. *"I was so honored to be a decorated combat veteran. I'm proud of the fact for one reason—the*

123

designation on the medals I was awarded signifies that I saved lives."

Among his decorations are the Purple Heart and the Bronze Star, recognizing his bravery and sacrifices.

Today, Sergeant Batista continues to serve the veteran community through his involvement in the Boston Wounded Veteran Run and other initiatives aimed at improving the lives of veterans.

Sergeant Richard Batista, we honor your service and sacrifice. Thank you for your unwavering commitment to our nation and for making a difference both during and after your time in uniform.

Welcome Home.

Seaman Curtis Beal

United States Navy (1969-1971)

Seaman Curtis Beal's service in the United States Navy during the Vietnam War placed him in one of the most dangerous and unforgiving environments of the conflict—the treacherous waterways of Cambodia. Serving as a Boatswain's Mate 1st Class, he was part of a River Assault Squadron, tasked with patrolling the narrow rivers and canals that served as critical supply routes for the Viet Cong (VC) and North Vietnamese Army (NVA).

Born and raised in Seattle, Washington, Beal was drafted into the Navy after graduating from high school. He completed basic training at Naval Station Great Lakes near North Chicago, Illinois, before being deployed to Southeast Asia. His assignment to a River Assault Squadron brought him to the frontlines of the war in Cambodia, patrolling the Mekong River and its labyrinth of waterways that crossed the border into Vietnam.

The "Brown Water Navy," as it came to be known, played a vital role in disrupting enemy supply lines and preventing troop movements along the rivers. However, these missions were fraught with danger. The dense jungle provided perfect cover for ambushes, and the small patrol boats left little room for defense. *"We did lots of patrolling along the Mekong River. Had to be on your toes constantly. SOB VC would ambush you in a second,"* **Beal recalled. "We were sitting ducks every time we went out."**

The patrols were grueling and unpredictable. *"It was 15 minutes of absolute hell, with an hour of frustration and the rest of your day wondering if you were going to be alive when you went out on your next mission,"* he said. The River Assault Squadrons operated under constant threat from hidden snipers, booby traps, and enemy boats armed with machine guns. The VC knew the

125

waterways intimately, using them to transport supplies and evade detection.

Despite the dangers, Seaman Beal formed close bonds with his fellow sailors. *"We had a really close-knit group—guys that had your back no matter what,"* he said. He shared a harrowing story of an ambush in which their boat was hit by an explosion. *"Our Petty Officer got thrown over the side. We yanked him out right before a huge goddamn alligator got him—all this while we were taking on fire. It was madness."*

The chaos of river warfare left little room for rest or celebration. When asked about entertainment and holidays, Beal responded emphatically, *"NO, NO NOTHING! Not a goddamn thing. We were trying to stay alive. Nobody, I mean nobody, thought of entertainment or Christmas when we were on patrol."* The brutal reality of war was ever-present, and Beal carried the weight of loss and trauma. He recalled a close friend who returned home in a wheelchair and tragically took his own life. *"Goddamn shame,"* he said, his voice filled with sorrow.

River patrols often required confronting massive enemy movements. Beal described seeing thousands of enemy soldiers attempting to cross the rivers. *"The enemy was hiding in the bushes, and here comes the VC down the trail—at least 1,000 men, for chrissakes. They had to get by us. Some nights they did, but most nights they didn't."*

The experience left deep emotional scars. When asked if he would ever consider returning to Vietnam or Cambodia, Beal's response was swift and resolute. *"NEVER! Too much pain and too many goddamn memories. None of them are any good."* He struck the table with his hand, signaling the end of the conversation. *"THAT'S ENOUGH,"* he said, rising from his seat.

Seaman Curtis Beal's service in the River Assault Squadron exemplified his courage and resilience in the waterways of Vietnam and in 1970 he was awarded the Purple Heart.

Seaman Curtis Beal, thank you for your service to our great country.

Welcome Home.

Specialist 4 Combat Medic Bill Blaisdell

United States Army (1969–1970)

Bill Blaisdell served his country in the United States Army from 1969 to 1970 during the Vietnam War, taking on one of the most dangerous roles imaginable – combat medic. After completing basic training at Fort Lewis in Washington State, he was deployed to Vietnam as a 20-year-old and assigned to the 1st Infantry Division at a small base camp in Dau-Tieng.

Dau-Tieng, located in the heart of South Vietnam, was home to a forward base that provided support to units operating in the area, often in dense jungle terrain. This region was a hotspot for Viet Cong activity, with constant threats of ambushes, booby traps, and mortar attacks. As part of the famed 1st Infantry Division, also known as "The Big Red One," Blaisdell faced relentless combat conditions. Known for their grit and resilience, soldiers of the 1st Infantry Division endured intense firefights, often in close proximity to the enemy.

For Blaisdell, the first six months of his deployment were spent in the field under heavy and continuous combat, an experience he described as terrifying and chaotic. *"Most medics didn't make it through the first six months. I was scared and confused,"* he recalled. His baptism by fire came on his very first day, during a firefight that took the life of his trainer, forcing him to step into a leadership role immediately. *"It was chaos,"* he remembered, as he took charge of prioritizing treatments and helicopter evacuations for the wounded.

128

Combat medics like Blaisdell carried the weight of their comrades' lives on their shoulders. Tasked with providing first aid in life-threatening situations, often under enemy fire, they were both revered and highly targeted. The mental and physical toll was profound, with sleep deprivation and constant fear becoming part of everyday life. *"Anybody that wasn't afraid was either a psycho or a liar,"* he said.

After six grueling months in the field, Blaisdell was reassigned to a base camp at Lai-Khe, the headquarters of the 1st Infantry Division. Nicknamed "Dr. Delta," Lai-Khe housed an aid station equipped with four treatment tables in a tent and helicopters on standby to evacuate the seriously wounded. While less chaotic than the field, the work at Lai-Khe was no less harrowing. Blaisdell described the grim reality of the makeshift morgue, which filled up quickly during heavy combat. One haunting memory involved a moment when he thought he had mistakenly placed a live soldier in the morgue. Although his commanding officer reassured him that post-mortem movements were possible, the experience stayed with him, leading to years of self-doubt and reflection.

Blaisdell's skill as a medic was recognized by his commanding officer, Dr. Weatherby, who praised his expertise in lifesaving procedures like inserting chest tubes. Despite his achievements, Blaisdell remained deeply affected by the loss of those he treated. His humility and dedication earned him the Bronze Star, a decoration awarded for acts of heroism, and merit. Blaisdell also suffered wounds during his service but chose not to pursue a Purple Heart.

Looking back on his service, Blaisdell reflected on how the Army shaped his life. *"The Army really straightened me out. It changed the path I was going down and provided a great foundation for my success in civilian life,"* he said. After his military career, he earned a degree from Northeastern University and became a major success in the telecommunications industry. Today, he continues to serve his community through his roles on the Cape & Islands Veterans Outreach Center board, the Mashpee Zoning Board of Appeals, and as a manager at New Seabury Country Club in Massachusetts.

Combat Medic Bill Blaisdell, thank you for your service to our great country and welcome home.

Corporal John "Jack" Demello Jr.

United States Army (1967 – 1969)

It was an absolute pleasure to interview one of my favorite people of all time for Veteran's Spotlight, Corporal John Demello Jr., who served his country in the United States Army from 1967-1969. Not only is this gentleman a lifelong friend, but he also served as a solid mentor early in my life. A proud graduate of Lawrence High School, where he was a star athlete, he later earned a degree from Bridgewater State College and went on to earn his master's degree at The University of Massachusetts.

Corporal Demello completed his basic training at Fort Dix, NJ, followed by Advanced Infantry Training at Fort Jackson, SC. Fort Jackson, located in Columbia, South Carolina, was one of the main training centers for soldiers heading to Vietnam during the war. Advanced Infantry Training (AIT) was intense and physically demanding, designed to prepare soldiers for the brutal realities of jungle warfare. Infantrymen underwent rigorous drills that included, combat tactics, weapons training, close-quarters combat, guerrilla warfare and live fire drills.

He recalls a sobering moment from training: *"The worst speech I've ever heard was from my 1st Sergeant. He said, 'Some of you guys are going to get wounded, and some of you are going to get killed.' I thought, this is getting better all the time!"* he said with a laugh.

131

When he arrived in Vietnam, he landed during daylight, a moment filled with nerves and uncertainty. One of the first people he bonded with was a fellow soldier from Taunton, last name Dull. *"We were in the same room in Advanced Infantry Training, we were in the same Battalion, Division, Company, and Platoon... he was killed during a patrol."* When asked about his reaction to his friend's death, Corporal Demello was visibly emotional. *"I was broken because I wasn't with him. We always spoke of getting in touch with each other's family if something happened to one of us and I wasn't able to after my discharge. Things happen, and life goes on. But fifty years later, I received info and found his three sisters, went to his grave too, and whole thing helped to bring a bit of closure."*

While in Vietnam, Corporal Demello spent time at a base camp in Cu Chi. *"Most of our time, we were in the field as we never stayed in one place too long, and we covered quite a bit of territory."*

One of the most dangerous aspects of his time in Vietnam was patrolling the Ho Chi Minh Trail, a vast network of jungle paths and roads used by the North Vietnamese Army to transport troops and supplies from North to South Vietnam. The trail, which wound through Laos and Cambodia, was heavily bombed by U.S. forces, yet it remained a lifeline for enemy operations.

He shared a particularly eerie memory from one of those patrols. *"I was on the Ho Chi Minh Trail in the middle of the night. It was so dark, you couldn't see your hand in front of your face. Suddenly, I was walking and came upon two headlights shining right at me. I stopped, took cover, and radioed headquarters. They said to stand by. Several minutes later, I received a message to disregard. I still don't know what the heck that was all about."*

Despite the dangers and chaos of war, there were moments of unexpected humanity. Relationships between U.S. troops and

132

Vietnamese villagers were complicated—often shaped by fear, mistrust, and occasional acts of kindness. While official policy discouraged fraternization, soldiers frequently found themselves interacting with locals. Some villagers were friendly, others were caught in the crossfire of a war they never asked for. Corporal Demello recalled a humorous story from one such encounter:

"We had set up outside a village. You're not supposed to fraternize with the villagers in this situation. A couple of days prior, I had shaved my head and was sitting there with my troops. I noticed this young girl kept staring at me. Then she got up and started walking toward me, which was a big no-no. She stopped, then kept walking again. Everyone is now looking at the girl. She comes right up to me, puts her hand on my head, and says, 'BUDDHA!'" he said, chuckling at the memory.

When asked about the holidays, he reflected on the sense of loneliness but also the deep friendships forged in war. *"You felt alone, but you shared it with close friends. I wrote my parents every third day. Never talked about anything real that was happening. I didn't talk about anything; 'we've been getting lots of rain... the Water Buffalo are great'... just ragtime stuff. My father wrote me a note demanding to know what was really going on. So, I wrote him and told him. He never wrote back. I think he didn't want to know the danger I was in."*

May of 1968 was the deadliest month of the entire Vietnam War for U.S. forces, with 2,169 American casualties. *"We lost about half of our battalion, which was extremely sad."* He recalled his first experience in battle: *"There's fear initially, but then the adrenaline takes over."* When asked if he was ever afraid, he gave a simple one-word response: *"Sure."*

Like many who served, Corporal Demello has been a champion for Vietnam veterans suffering from PTSD. The psychological wounds of war ran deep, and he witnessed its impact firsthand. *"We were being mortared in a bunker. All of a sudden, one of our guys snaps and starts to run out to attack the enemy. We had to tackle him and hold him down. He was wounded seven different times. Mentally, he was just a wreck."*

When he returned home, Corporal Demello, like many Vietnam veterans, struggled to talk about his experiences. *"When I came back, I didn't talk about my service in Vietnam at all, didn't make any contact with anyone. Then a guy called me from my battalion, then another guy. This was the soldier who carried me to my helicopter when I was wounded. I lost track of him. He had been in a VA Psychiatric Hospital since the war. Thankfully, he's been discharged. I talk with him periodically."* He paused, holding back emotion.

Vietnam veterans faced a difficult homecoming. Unlike soldiers from previous wars, many returning Vietnam veterans were met with hostility and protests rather than gratitude. Though Corporal Demello holds no ill will toward war protestors, one thing still weighs heavily on him. *"When they called us baby killers. That hurt a lot,"* he said softly.

Reflecting on his service, he expressed deep pride: *"I was proud of what we all did. I suppressed it for so long... proudest thing I ever did."* Though his humility prevents him from elaborating much on his commendations, he is particularly proud of three, The Purple Heart, The Combat Infantry Badge and The Service Medal with 2 Bronze Stars.

Corporal John Demello Jr., thank you for your service to our great country, and welcome home.

Specialist 4 Dave DiMestico

United States Army (1970-1971)

It is an honor to feature Specialist 4 Dave DiMestico, who served his country with courage and distinction in the Vietnam War. A graduate of Lawrence High School, Dave attended college before being drafted in 1970 alongside his friend, Steve Paltz, a fellow Falmouth, MA native. Basic training began that summer at Fort Dix, NJ. *"I was in good shape, really didn't find it hard but it was summer and it was hot, took advice from seasoned soldiers. They said, keep a low profile and don't let them know your name,"* Dave recalled, embracing the wisdom that would carry him through the challenging months ahead.

After basic training, Specialist DiMestico was sent to Advanced Infantry Training (AIT) at Fort Lewis, WA. Fort Lewis, located near Tacoma, was renowned for its rigorous field exercises and unpredictable weather. Dave described spending little time in the barracks, instead camping in the field under tents while learning wartime strategies and mastering weaponry. *"It rained during the day, cleared at night, and the stars would come out,"* he reminisced, capturing the duality of hardship and natural beauty that defined his training experience.

Next came Airborne School at Fort Benning, GA, a legendary site known for its intense paratrooper training. Soldiers here learned to jump from planes, executing the precise and dangerous maneuvers required of the airborne infantry. After earning his jump wings, Dave was sent to Vietnam's Central Highlands, above Cam Ranh Bay, for two weeks of jungle warfare training. This region, with its dense foliage and rugged terrain, was a strategic hub for U.S. forces, and it prepared soldiers for the challenges of guerrilla warfare in Vietnam's unforgiving landscapes.

In Vietnam, Dave joined the 173rd Airborne Brigade, the "Sky Soldiers," a storied unit with a reputation for bravery and tenacity. Assigned to an infantry platoon, Dave quickly rose to the role of squad leader, managing a team equipped with M-60 machine guns, grenade launchers, and ammunition carriers. At 22 years old, he was one of the older soldiers in his unit, a factor that helped him take on leadership. *"I adapted fairly easily and ran it like, 'Everyone do their job, and we won't have a problem,'"* he reflected. In quieter moments, he carried a simple book in his pocket, a small comfort amidst the chaos of war.

Dave's first firefight remains vividly etched in his memory. *"We were on helicopters, sitting on the edge. The door gunner taps you when it's time to jump,"* he explained. A miscue led him to leap prematurely, landing roughly 10 feet below. *"Definitely a little scary,"* he admitted. Moments like these, combined with the ever-present danger of combat, underscored the intensity of his deployment. He described a harrowing experience during a firefight where his unit was nearly overrun. *"They called in an airstrike,"* he said, recounting the sensation of being lifted off the ground by the force of 500-pound bombs. *"I was praying I'd make it through."*

Another vivid memory was the constant threat of snakes in the jungle. *"I hate snakes,"* he emphasized, describing tunnels of vines above that served as *"highways"* for the reptiles. Among the most dangerous were venomous species like the Malayan pit viper, known for its aggressive nature and potentially fatal bite, the highly venomous cobra, and pythons. These snakes were not just a threat to safety but also added to the psychological strain of navigating the hostile terrain.

Holidays in the field were a somber affair. *"No Bob Hope shows for us,"* he said. After weeks in the field, brief respites included warm beer and as much steak as one could eat—rare moments of comfort in a harsh environment.

Returning home brought its own challenges. The Vietnam War was a divisive conflict, and many veterans faced protests and a lack of recognition. *"I was proud to serve, but I thought Vietnam was a waste of time,"* Dave said candidly. The sight of young men returning with severe injuries fueled his frustration, as did the hostility from some civilians. *"I had some animosity toward the protesters,"* he admitted. Like many veterans, Dave chose not to discuss his experiences publicly, carrying his memories quietly.

Service runs deep in Dave's family. His father, Paul, served in the Army Air Force during World War II, his brother Pete was a Marine during the Vietnam era, and his son Mike has served in the Navy for 13 years.

Specialist 4 Dave DiMestico, we thank you for your service to our great country and welcome home.

Sergeant Al DiMuzio

United States Army (1967-1970)

Al DiMuzio served his country in the United States Army as a Sergeant from 1967 to 1970. Born in Newton, Massachusetts, he grew up in Framingham and enlisted in the Army shortly after graduating high school. Reflecting on his decision, he shared, *"I was an air-headed 18-year-old who had no idea what I wanted to do. My motivation for volunteering came from my family. My grandparents were immigrants, and my dad and uncle both served in World War II. Their teachings reinforced my desire to serve my country. I thought, 'My country needs something,' and I wanted to answer that call."*

After enlisting, Sergeant DiMuzio completed basic training at Fort Jackson, South Carolina, before advancing to Fort Leonard Wood, Missouri, for specialized training. *"It was the next level of training where I learned to operate heavy machinery,"* he explained. *"It was fun and interesting because I came from a family business in sand, gravel, and concrete. It felt like I was right at home."*

His first major assignment was at Fort Sill, Oklahoma, where he worked as a crane operator. In 1969, at just 20 years old, Sergeant DiMuzio was deployed to Vietnam. He landed in Long Binh and Bien Hoa, where his initial impressions of the environment were stark and unforgettable. *"It was sort of like a global physical experience,"* he said. *"It was hot, humid, and it stunk beyond belief with smells you don't recognize. I knew I wasn't home."*

Holidays during his deployment were challenging. *"I experienced them, but they were so non-consequential. Stepping off that plane made it clear this was not home. My family wasn't there,"* he shared. One bright spot, however, was receiving care packages.

138

"Those packages created an instant connection to my family. They sent all the Italian delicacies—cheese, salami, pepperoni— all the things from home. It reminded me of my roots and kept those strings of connection alive."

When asked about mentorship during his service, Sergeant DiMuzio explained that while he did not have a formal mentor, he valued building connections. *"I didn't mentor, but I always extended an invitation for friendship. Friendship can be a form of mentoring,"* he reflected.

Sergeant DiMuzio described the camaraderie among his fellow soldiers as a positive and unifying experience. *"It wasn't 100% consistent, but I had the opportunity to get to know people of different religions, beliefs, backgrounds, and cultures, and to embrace that diversity. The camaraderie and commitment were always there. We'd get mortared and shelled, but the bond with my fellow soldiers kept us going."*

However, his time in Vietnam was not without its darker moments. Sergeant DiMuzio became emotional when discussing the chemical exposure to Agent Orange. *"Every Monday morning, we had formation. A C-130 would fly over and drench us with the spray from Agent Orange. This happened weekly. My eyes would burn, it was on my face, my lips, and I'd breathe it in. We had no safety equipment, and as a result, I now have persistent cardiac issues. It's my general belief that the 'higher authority' didn't, and perhaps still doesn't, believe or trust us when we spoke about the effects."*

He also shared his perspective on the anti-war protests of the time. *"My first exposure to them was before I enlisted, and I understood their perspective. But once I was deployed, the protests became a conflict within themselves. My experience was that they had no respect for the military. When I returned home, I remember walking through the airport, and someone spit at my feet. That person had no idea what I had gone through. My thought was, '(Expletive) you, pal.' Despite that, I was grateful to be home and confident I had fulfilled my duties."*

Reflecting on his military service, Sergeant DiMuzio expressed a mix of pride and understanding of its lasting impacts. *"I took pride in my service, my duties, and what I accomplished. I feel successful in what I did, but there's no way to experience that without going through some form of trauma. It became a deeper reality for me."*

Today, Sergeant DiMuzio resides in Barnstable, Massachusetts. His story is a testament to the sacrifices, resilience, and strength of those who serve.

Sergeant Al DiMuzio, thank you for your service to our great country, and welcome home.

Colonel Jack Donovan

United States Marine Corps (1964-1966)

Colonel Jack Donovan served two years in the United States Marine Corps and 24 years in the United States Army. His assignments took him across the globe, including pivotal moments during and after the Vietnam War. A humble man with a wealth of stories, Donovan chose to highlight the bravery and humanity of his fellow soldiers, particularly in their interactions with Vietnamese children and efforts to account for missing American servicemen.

In Vietnam, Colonel Donovan witnessed firsthand the complicated relationship between American soldiers and the local population. Despite the hardships and horrors of war, many soldiers formed bonds with Vietnamese children, offering them kindness in small but meaningful ways. *"There were two street kids, orphans, that I would see all the time,"* he recalled. *"They sold gum and candy. One of my friends, Waters, took a liking to them. The kids called him 'Ong Nuoc' (Mr. Waters)."* Waters wanted to buy the kids a bicycle, but they declined, preferring to use the money to pay for their education at a local orphanage. Donovan, in turn, took them shopping for clothes.

However, these interactions were not without risk. The Vietnamese government viewed any positive relationship between locals and Americans with suspicion, often resulting in arrests. *"I found out the government had arrested the kids because they got too close to us,"* Donovan said, his voice filled with resolve. *"I used my connections to get in front of some very influential people and told them straight out that this will never happen again. I said if*

141

anything happened to those kids, I would come back and take appropriate action." His icy stare made it clear that he wasn't bluffing.

In the A Shau Valley, a place of heavy combat during the war, Donovan witnessed another example of soldiers connecting with children. *"One of my soldiers, Mike Eagle, began trying to converse with the local kids. He gave them candy and started singing nursery rhymes like 'Old MacDonald.' More kids began showing up every night to listen. But then, one night, they stopped coming. The Vietnamese government shut it down. They never wanted Americans to be seen in a positive light,"* he recalled with sadness.

Donovan's commitment to protecting children extended to teaching them about the dangers of unexploded ordnance. *"I brought a bomb ordinance man with me to educate the kids on how dangerous bombs and explosives were. We wanted them to stay away,"* he explained. Tragically, one day, a young boy was killed after playing with a live bomb that detonated. *"We ran to the site after hearing the explosion. It was heartbreaking."*

Donovan's work in Vietnam extended to the recovery of American remains. His missions often involved digging and sifting through crash sites to find traces of missing soldiers. He shared one particularly emotional story about a young American man who joined his team in search of his father, whose plane had gone down in Vietnam. *"I prepared him as best I could. I told him it was going to be an emotional day and that if we found a bone fragment or a piece of uniform, it might not necessarily be his father. I told him we could take breaks to pray, talk, cry— whatever he needed. It was a really emotional day for me personally. I told him that when he returned to the United States, he needed to tell people what our guys are doing over here, what these guys are like that are searching for his dad."*

142

Colonel Donovan also shared a powerful story passed down from a fellow soldier about Admiral John McCain, the father of Senator John McCain. Admiral McCain, as Commander of Pacific Command, received a briefing at the Pentagon about American POWs in a Vietnamese prison camp. He asked the officer giving the briefing, *"Is my son Johnny in that camp?"* The officer replied no. Admiral McCain's response? *"Let's go in anyway."* For him, it was never just about his son—it was about saving as many American lives as possible.

The plight of American POWs in Vietnam remains one of the most painful aspects of the war. Thousands of U.S. servicemen were captured during the conflict, often enduring years of brutal treatment and torture in camps such as the infamous "Hanoi Hilton." Colonel Donovan's involvement in POW/MIA recovery efforts demonstrated his unwavering commitment to bringing every American home, dead or alive.

When asked about his long military career, Colonel Donovan responded with deep appreciation. *"With one minor exception, I loved every minute of it. I was in special units with great people."* His service exemplifies the courage, compassion, and determination of the men and women who served in Vietnam.

Colonel Jack Donovan, thank you for your service to our great country.

Welcome Home.

Colonel Kevin Doyle

United States Marine Corps (1967-1997)

After interviewing this veteran, I came away deeply impressed with his character and, most importantly, his humility. Colonel Kevin Doyle spent 30 years in the United States Marine Corps, serving in both active and reserve capacities. Military service ran deep in his family—his father was a highly decorated Navy Lt. Commander during World War II, and his son Brian served 23 years in the Army. Colonel Doyle grew up in Arlington, Massachusetts, attended Arlington High School, and in 1963 received a Navy ROTC Scholarship to Holy Cross, where he was commissioned as a Marine officer in 1967.

While attending Vietnamese Language School in Washington, D.C., Colonel Doyle witnessed firsthand the raw emotions that followed the assassination of Dr. Martin Luther King Jr. on April 4, 1968. *"It was eerie. The whole sky darkened, and the air was thick with smoke from burning buildings and belching buses—it felt like something out of a Hollywood production,"* he recalled. The city was on edge, with riots erupting in the streets, and fear hung heavy in the air. *"A bunch of us were leaving school, and I said we should all grab a cab,"* he continued. But an African American Staff Sergeant named Anderson shook his head and said, *'No sir, I don't think I should do that today.'"* That moment stuck with Colonel Doyle—a stark reminder of the deep divisions and pain the nation was experiencing.

Later that year, he deployed to Vietnam as an advisor to the Regional Forces, where he also commanded troops in combat. Fighting through the dense jungle and flooded rice paddies was

144

grueling, and the A Shau Valley—one of the most treacherous areas in Vietnam—was particularly unforgiving. *"You could smell the damp earth, hear the rustling in the jungle at night— sometimes it was the wind, sometimes it wasn't,"* he recounted. The valley was a critical supply route for the North Vietnamese Army, making it the site of intense battles. Patrols were highly dangerous, and for six months, he slogged through the rice paddies, knee-deep in mud, leeches clinging to his skin, all while constantly on high alert for ambushes. He also spent 30 harrowing days navigating the infamous Ho Chi Minh Trail, where enemy forces lurked in the thick underbrush, waiting for the right moment to strike.

One of the most powerful memories he shared was of a soldier who had been gravely injured in battle, losing part of his face in an explosion. Colonel Doyle stayed with him through the night, offering comfort and reassurance as they waited for evacuation. When the helicopter finally arrived at first light, a sling was lowered to lift the wounded man to safety. As he ascended into the sky, he raised a weak but determined hand and saluted Colonel Doyle. Even decades later, the memory still moves him.

Fast forward 25 years to the Vietnam Veterans Memorial Wall in Washington, D.C. Colonel Doyle had gone to pay his respects when he noticed a familiar face among the crowd. Their eyes met. Recognition flickered. They began talking. After more than two decades, Colonel Doyle was reunited with the Marine he had watched over that night in the jungle—Rick Krets. *"Do you know how many people I told about your courage?"* Colonel Doyle exclaimed. Krets, his voice filled with emotion, responded, *"Sir, do you know how many people I've told about your inspiration?"*

The Vietnam Veterans Memorial Wall stands as a solemn tribute to the 58,000 Americans who lost their lives in the war. The polished black granite reflects the faces of visitors as they trace their fingers over the engraved names of the fallen. It is a place of remembrance, of healing, and for many, of reunion. For Colonel Doyle, it was the site of an unexpected and deeply moving moment—proof that bonds forged in war never fade.

When asked about a mentor during his service, Colonel Doyle answered without hesitation, *"P.K. Van Riper. He went on to become a three-star general. He taught composure on the battlefield, how to conduct yourself without losing your mind— he was a proud Marine."* He also made a point of recognizing their mutual friend, Tommy Leonard, as a source of inspiration. *"Tommy served in the Marines for four years but has carried the tradition for 85,"* he said proudly.

I asked Colonel Doyle his opinion on several topics:

Leadership: *"You're driven by accomplishments and taking care of your people."*
Losing A Soldier: *"Devastating. You own it and carry it with you."*
Entertainment Overseas: *"Very important. The care packages, anonymous letters, and USO shows meant a great deal."*
POWs: *"Anytime you think you're having a bad day, talk to a POW. I've had the honor of knowing a few."*
Overall Military Experience: *"VERY positive. A beacon in my life. Never stops. My moral compass."*

Colonel Doyle's awards and medals are extensive, but in characteristic humility, he mentioned only a few: the Legion of Merit, two Navy Commendation Medals with "V" for Valor, and the Vietnamese Silver Star.

Colonel Kevin Doyle, thank you for your service to our great country.

Welcome Home.

Specialist E-4 Bob Dutra

United States Army (1969)

Specialist Bob Dutra served his country with dedication and resilience during the Vietnam War, completing a 10-month combat tour from February to December of 1969. A graduate of Lawrence High School, Bob initially joined the Army National Guard and began his military training at Fort Leonard Wood, Missouri. *"Basic training was fine. I was 18, young, in shape, and got to meet some good people,"* he recalled. After completing Advanced Infantry Training (AIT) and a brief stint at Fort Benning, Georgia, he shipped out to Vietnam.

The journey to Vietnam left an indelible impression on him. *"The plane ride was eerily quiet; nobody said a word. When we landed, the stewardesses hugged the soldiers and cried. I thought, 'What did I get myself into?'"* Once in Vietnam, Bob was assigned to the Central Highlands with the 1st and 14th Artillery and later to the 1st of the 96th Infantry of the American Division. His unit frequently moved ahead of artillery positions, landing in helicopters on freshly sprayed Agent Orange fields. *"The chopper blades never stopped. You jumped out, and all the sand, debris, and Agent Orange came right at you—into your face, eyes, mouth, and on your skin,"* he recalled. *"You always worried about snipers during those landings. That was the time to be scared."*

Life on the ground was grueling and unforgiving. *"Nighttime was the worst. It wasn't comfortable; you couldn't see anything, and the enemy could sneak up on you in a split second. You were always hot, sticky, and sweaty. Infantry guys had it the*

147

toughest—I had a lot of respect for those guys." Showers were rare luxuries. *"At base camp, they poked two holes in a large drum, filled it with water, and added faucets. That was our shower, but who knows what had been in those drums before?"*

Holidays during the war were particularly poignant. *"They didn't really bother me because the next guy was in the same boat. But I felt for the married guys. They'd get cassette tapes from home, go off by themselves, and listen to them. That was tough for them."* Bob shared a touching memory of receiving a care package from his father. "He sent linguica packed in Crisco inside a coffee can. I shared it with everyone and cooked it over a fire. It was much better than sea rations."

The experience of losing fellow soldiers left an emotional scar. *"It stuns you—like a shock. It's something you never really get over,"* he said quietly. Amidst the challenges, he found small moments of connection, like meeting another soldier from New Bedford. *"We talked a few times. It was one of the few bright spots."*

Reflecting on his service and the protests at home, Bob expressed mixed emotions. *"My fondest memory was getting my notice to come home. Nobody wanted to be there—all the casualties, the deaths, and then just leaving? The POWs were the real heroes to me, and the guys we left behind. I haven't forgiven the protesters to this day. Everyone's entitled to their opinion, but when you're there, you want to feel some support from the people back home."*

After returning from Vietnam, Bob continued to serve his community for over 32 years as a firefighter with the Falmouth Fire Department. He remains one of the most respected and well-liked members of his town.

Specialist Bob Dutra, a heartfelt welcome home and thank you for your service to our great country.

Captain John Ferguson Sr.

United States Air Force (1958–1969)

While interviewing this veteran two words stood out—class and humility. Captain John Ferguson Sr., who served in the United States Air Force from 1958 to 1969, grew up in Westerly, Rhode Island, and graduated from the University of New Hampshire. He entered the service as a Second Lieutenant through the ROTC program in college and retired as a Captain.

Initially stationed in the Philippines, Captain Ferguson's unit was later deployed to Dong Ha, Vietnam, as the U.S. sought to establish a radar site near the Demilitarized Zone (DMZ). Reflecting on this transition, he remarked, *"We had no idea what we were getting into… my first impression was to stay alive… didn't know one Vietnamese from the other."* He shared a harrowing story about a Vietnamese man who had obtained a security clearance to work in their kitchen. *"We didn't know it at the time, but he was actually a VC [Viet Cong] soldier… they shot him while he was lobbing grenades toward an area with Marines. That place was somewhere we never should have been. The only thing war does is kill people."*

Captain Ferguson credited the close bonds within his unit for their ability to endure the constant threats. *"We were mortared a few times under full moons… we had no casualties except equipment. That's because everyone worked together… we were close,"* he recalled. Regarding holidays during his deployment, he dismissed their significance. *"It just became one day… I would have liked to have been home with my wife and four kids, but I wasn't. I looked at it as just another day."*

When asked if he had the chance to see Bob Hope, Captain Ferguson responded with a touch of humor. *"He didn't dare come that close to the DMZ,"* he said, referring to Hope's well-known USO performances. Despite this, Bob Hope was renowned for his willingness to perform in perilous locations. His commitment to entertaining troops often took him to areas close to combat zones, although rarely as close as the DMZ in Vietnam. In one instance, Hope performed on an aircraft carrier under the cover of darkness and, on another occasion, at Long Binh near Saigon, just miles from active enemy forces.

Instead of Hope, Captain Ferguson recalled the visit of a two-star general, which he found more practical. *"I liked seeing him because he could get us equipment,"* he said with a smile.

Among his most vivid memories was a decisive moment involving a besieged unit at Khe Sanh. *"We had four fighter aircraft on standby under my command. Khe Sanh was under heavy mortar fire, and I had to go through three or four communication channels all the way back to the Pentagon for approval to send the fighters. I wasn't getting the go-ahead, but Khe Sanh radioed that they'd mark targets with colored smoke."* Taking initiative, he authorized his two lieutenants to deploy the aircraft, which successfully stopped the mortar attack, saving the unit from capture or death. *"The next day, a Lieutenant Colonel came in and ripped me... wanted to know who gave permission to send the fighters. I told him I did. He never said I did the right thing,"* Ferguson recalled.

Captain Ferguson emphasized the traits of a strong leader in combat: understanding one's responsibilities, ensuring the well-being of troops, and maintaining morale—even by ensuring the delivery of mail. When asked what he thought about most when coming home from Vietnam, he said, *"I wanted to see my family badly. We got mortared a few times, and fortunately, we didn't lose any men."*

Now 86, Captain Ferguson wants it noted that he has battled prostate cancer and Parkinson's disease due to exposure to Agent Orange.

Captain John Ferguson, thank you for your service to our great country.

Welcome Home.

Airman 1st Class Tim Flynn

United States Air Force (1963–1967)

Tim Flynn served his country in the United States Air Force with distinction as an Airman 1st Class from 1963 to 1967. Born in Newport, KY, Flynn enlisted at the age of 17, with his father co-signing his enlistment papers. ***"Had a couple of friends that were going in, so I decided that I wanted to as well,"*** he remembered. After basic training at Lackland Air Force Base in San Antonio, Texas, where the sweltering summer heat and salt pills were a constant challenge, Flynn began charting his military path.

Initially interested in becoming a pilot, Flynn discovered he was colorblind, which eliminated that option. Instead, he was offered a choice between working in a supply depot or serving in the Air Police. He chose Air Police and completed an eight-week training course at Lackland. His first assignment was at Otis Air Force Base on Cape Cod, where he became intrigued by the specialized role of sentry dog handlers. Volunteering for Sentry Dog School, he returned to Lackland for another eight weeks of intensive training, learning obedience techniques, relationship-building with his assigned dog, and managing the critical bond between handler and canine.

In September or October 1965, Flynn graduated from training with his dog, Rommel, a German Shepherd, and prepared for deployment to Vietnam. Their preparation included acclimating Rommel to gunfire, explosions, and other battlefield conditions. Flynn and Rommel, along with about 50 other handlers and their dogs, departed for Vietnam on a C-130 cargo plane, traveling through bases like Hamilton AFB in California, Hickam AFB in Hawaii, and even stopping on Wake Island for emergency repairs. On the remote island, the handlers allowed the dogs to run on the beach, a rare moment of respite in their journey.

In Vietnam, Flynn and Rommel were tasked with patrolling the perimeters of airstrips and airports, safeguarding areas with highly volatile materials such as gas and oil. The job was nerve-racking and dangerous. *"I was afraid every (expletive) night. When your dog alerts, he's not playing games—something's out there, human or animal,"* Flynn recalled. Sentry dogs were invaluable in detecting hidden threats, whether from enemy forces or wildlife. Flynn shared a harrowing account of hearing reports of a tiger on the prowl while patrolling a bomb dump. *"I'm thinking, 'Jesus Christ!' Thankfully, we never ran into the tiger, but sundown still affects me to this day. I still get very anxious,"* he admitted.

Military working dogs like Rommel played a critical role in Vietnam, serving as scouts, sentries, and trackers. They were credited with saving countless lives by detecting ambushes, tripwires, and enemy movement. By the end of the Vietnam War, it is estimated that nearly 4,000 military working dogs served alongside U.S. forces, often in harsh and dangerous conditions. Tragically, most of these dogs were not brought back to the United States after the war due to outdated policies classifying them as "equipment."

Leaving Rommel behind was one of the hardest moments of Flynn's service. *"It was tough. You never want to leave your dog. We went through a lot together, and I didn't want to leave him. It was tough to do,"* he said, his voice heavy with emotion. Flynn later discovered that Rommel was assigned to another handler for about a year and a half, but beyond that, his fate remains unknown.

Reflecting on his service, Flynn expressed a mix of pride and disillusionment. *"Had a bit of satisfaction that I served my country. Glad I served. Saw a lot of corruption. Anyone that served in Vietnam, I support 100%,"* he said. Flynn's legacy includes not only his service but also his role as a devoted father to two daughters, Shannon and Tieraney.

Airman 1st Class Tim Flynn, thank you for your service to our great country and welcome home.

Sergeant Barry Funfar

United States Marine Corps (1966–1970)

Sergeant Barry Funfar served his country with distinction during the Vietnam War, flying an extraordinary 127 missions as a door gunner aboard the iconic Huey helicopter. Growing up on a farm in Lidgerwood, North Dakota, Barry learned the value of hard work from an early age. *"I worked 10-hour days with no overtime for $3.00, then went to the University of North Dakota, working at the Student Union for 90 cents an hour. My four brothers and I all did well. I tried to set a good example for them—guess you could say we were an American success story,"* he reflected with pride.

After basic training at Camp Pendleton in San Diego, Barry was assigned to the VMO-2 Marine Corps Air Wing at Marble Mountain Air Facility, just three miles from Da Nang. From February 1968 to October 1969, he served at what was infamously known as a "hot base." *"We got mortared and shot at constantly,"* he recalled. Initially, he worked with local Vietnamese civilians reinforcing bunkers and spoke highly of their dedication. *"I had a tremendous amount of respect for them."*

Sergeant Funfar's missions included night flying, medevacs, extractions, and providing critical air support to ground troops. He developed a deep empathy for the "ground pounders," the soldiers fighting on the ground. *"They loved seeing us come to get them. The life expectancy of a door gunner in combat was just five minutes, but you didn't think about it. Adrenaline takes over, and you do what you're trained to do,"* he said. Despite the constant

156

danger, Barry described the unique beauty of Vietnam's landscape, even amidst the devastation of napalm and bomb craters. *"I loved to fly. But when you see .51-caliber shells coming at you the size of footballs, that's when reality sets in."*

Entertainment and brief moments of joy were rare but treasured. Barry remembered Christmas of 1968 fondly. *"I volunteered to work—it was just another day for me—but I'll never forget Bob Hope's show. They set up speakers, and the applause and laughter echoed across the base. Bob Hope stayed to talk with us after the show. I have the greatest respect for that man."*

Life at Marble Mountain required constant readiness. *"There was a lot of talk about drugs and alcohol in Vietnam, but not in my unit. Marines couldn't even have hard liquor. If you were on flight duty, there was no way you could be on drugs. In my downtime, I played chess, read, and even played handball. You always had to be ready."*

Reflecting on his mentors, Barry immediately named Gunnery Sergeant Black. *"He was a great mentor. I think of him often—he always tried to dissuade me from flying."*

Coming home from the war proved deeply challenging. *"When I was discharged, they flew us into San Francisco and told us not to wear our uniforms. That was devastating—absolutely devastating."* The protests and insults he faced upon his return hurt more than the combat itself. *"I was filled with anger and depression. I didn't talk about it to anyone. It took years, therapy, and even trips back to Vietnam to begin healing. But the war never goes away. For those of us who were there, it never leaves."*

Despite the pain, Barry expressed pride in his service. *"I would do it again and again. I was proud of the job I did."*

Sergeant Barry Funfar, thank you for your extraordinary service to our great country. Welcome home and thank you for sharing your story.

Lieutenant Colonel John Harris

United States Air Force (1966-2000)

An absolute delight to interview, Veteran Spotlight Lieutenant Colonel John Harris, whose genuine, courteous, and down-to-earth manner made for a truly engaging conversation. Lt. Colonel Harris served his country for 34 years in the United States Air Force, Air Force Reserves, and Army National Guard. Growing up in the suburbs of Atlanta, GA, he graduated from the University of Georgia and was commissioned through ROTC in the Honor Guard. After completing pilot training at Moody AFB in Georgia in 1966, Lt. Colonel Harris was sent to Vietnam, where he was stationed at Ubon Royal Thai Air Force Base in Thailand with the prestigious 555th Tactical Fighter Squadron, 8th Tactical Fighter Wing. During his eight months there, he completed 100 highly dangerous air missions, risking his life in the skies over hostile territory.

When asked about the mindset of a pilot on such perilous missions, Lt. Colonel Harris shared, *"You have to be on your toes and extremely well prepared… if you're not scared, there's something wrong with you."* That fear, however, was never crippling. It became a driving force, sharpening his focus and reaction times in the face of constant danger.

Lt. Colonel Harris flew the formidable F4-C and F4-D Phantom II fighter-bombers. These planes were cutting-edge machines of their time, capable of flying at supersonic speeds and carrying a wide array of weaponry, including bombs, missiles, and guns. *"Our mission was to go in and bomb, then hang around to protect our strike force from the enemy MiGs,"* he explained. The F4-C was a

159

workhorse of the Vietnam War, known for its speed and versatility, but flying it over enemy territory wasn't for the faint of heart.

The threat from North Vietnamese MiGs was always present. These sleek, nimble Soviet-made fighters often darted out of the clouds or came up from hidden bases to intercept American strike forces. Lt. Colonel Harris had to maintain a constant state of vigilance. Seeing a MiG shoot down a fellow pilot was a haunting experience. *"We were on a bombing mission... there were surface-to-air missiles and gunfire coming right at us,"* he recalled. *"A missile took out the guy next to me. I saw his plane explode and then spiral to the ground... just a ball of fire going down. You never get over that."*

The skies over Vietnam were a chaotic, dangerous place. Surface-to-air missiles (SAMs), anti-aircraft artillery, and enemy planes created an almost apocalyptic environment during his missions. Harris described how, at times, the air would be filled with black smoke trails from missiles and explosions. *"It was like flying through a storm of fire,"* he shared.

Despite the ever-present danger, there were moments of levity. One Christmas, while preparing for a mission in Bangkok, Lt. Colonel Harris was summoned by Colonel Chappy James, a legendary figure in the Air Force, and asked to play Santa for the troops. *"Colonel James got me stuff for presents—cigarettes, candy, beers. I got a bomb cart and dressed it like a sleigh, and I found four of the smallest guys in the squadron to be my elves,"* he chuckled. The event brought joy to the base, though it almost got an Australian officer court-martialed when the Secretary of the Air Force unexpectedly arrived during their foam-snow celebration.

When asked about role models, Lt. Colonel Harris didn't hesitate to name Colonel Robin Olds, his Wing Commander. *"Fearless, smart... a natural pilot. He was an ace pilot in WWII and also an*

All-American football player at West Point. In my opinion, he's the best air combat leader in the history of our country."

Today, Lt. Colonel Harris lives in Mashpee, Massachusetts with his wife Susie, herself a veteran with 27 years of service in the Air Force and Army National Guard. A member of the Ancient & Honorable Artillery Company of Massachusetts, he reflects on his long career with humility. *"I enjoyed almost every minute of it."*

Lieutenant Colonel John Harris, thank you and your wife Susie for your service to our great country.

Welcome Home.

Captain Carter Hunt

United States Marine Corps (1969–1970)

Captain Carter Hunt served his country with valor and resilience during the Vietnam War, retiring as a distinguished officer in the United States Marine Corps. A native of Leominster, Massachusetts, Captain Hunt attended Boston College, where he excelled as a starting defensive end for the football team and graduated in 1968. Soon after, he entered Officer Candidate School (OCS) in Quantico, Virginia, graduating in April of 1969, followed by completing Basic School in September. By October, he was deployed to Vietnam.

Arriving in a war-torn region, Captain Hunt quickly realized the gravity of the situation. *"It was a busy place,"* he recalled. *"We were scrambling to headquarters to get our assignments."* Assigned to Echo Company, 5th Marine Division as a Rifle Platoon Commander, he joined his platoon at Hill 52, where he met his seasoned Platoon Sergeant and began working alongside Vietnamese regional forces. He described those early days as a mix of routine patrols through rice paddies and sporadic enemy contact. *"We moved to the Arizona Territory, where enemy engagements became much more frequent,"* he explained. In the chaos of combat, trust was paramount. *"You have to depend on your squad leaders. My platoon had been together for a while, so we had that experience in combat."*

Captain Hunt also shared memories of his first holiday in Vietnam. *"Thanksgiving was my first. We got menus on a piece of paper, turkey and mashed potatoes in vacuum-sealed cans, a can of beer, and a can of soda. There was a lot of bartering with guys who drank and those who didn't. We even got a cake leftover*

162

from the previous Marine Corps Birthday," he recounted with a chuckle.

However, war often brought heartbreak. Reflecting on a dark day at Son Vu Gai, he recalled losing two Marines to a grenade that exploded prematurely. *"There's no way to explain it, losing a Marine. You're living cheek to jaw with each other; you're an integral part of a team. It happens in combat, but you have to keep going with your mission."*

One particularly harrowing memory came from patrolling Go Noi Island. *"We got news we were going to be ambushed. My lead squad heard activity and got into a firefight. The next morning, we found two VC bodies with AKs and several magazines. Later, we moved to a new position on the island and came under fire— only to realize it was our own troops shooting at us. Two of my sergeants were killed, and three other Marines were wounded. We found out that our location wasn't marked on the headquarters maps,"* he said somberly.

Despite the challenges and tragedies, Captain Hunt reflected on the honor of serving his country. *"Duty, honor, country—those values were instilled in me from the beginning. My father was a WWII veteran, my grandfather fought in WWI and the Spanish-American War, and my great-grandfathers served in the Civil War. If I was going to fight, I wanted to fight with people who knew how."*

When asked about the anti-war protests back home, he responded thoughtfully. *"That's what we were defending—the right to free speech, religion, and more. What bothers me is the people who went to Canada and didn't serve in any capacity. I believe in universal service—whether it's the military, Peace Corps, nursing homes, something."*

As a Marine, Captain Hunt described the bond forged in combat. *"It's a special feeling. We're a band of brothers, all going through the same experiences, enduring the tough times, and wanting to do the right thing."*

Captain Hunt was awarded the Purple Heart for his injuries sustained in Vietnam. His service embodies the highest ideals of duty and honor, and his story reminds us of the sacrifices made by so many during the Vietnam War.

Captain Carter Hunt, welcome home, and thank you for your extraordinary service to our country.

Dr. Jim Kenner

United States Army (1967-1970)

Dr. Jim Kenner served as a trauma surgeon in the United States Army during one of the most intense periods of the Vietnam War. Born in Montgomery, Alabama, he completed his medical training at the University of Alabama and entered the Army through the Berry Plan, a Defense Department program allowing doctors to finish their education before fulfilling military service obligations.

In March 1967, Dr. Kenner arrived at Tan Son Nhut Air Force Base near Saigon, a major hub for wounded soldiers evacuated from the frontlines. The transition from medical school to a war zone was jarring. *"The smell was the first thing that got me,"* he recalled. *"A horrible, rancid odor—honestly, it just smelled like death everywhere I went. It hit you like a ton of bricks."*

Tan Son Nhut's medical facilities were constantly overwhelmed. Dr. Kenner's days began at 6:30 a.m. and often stretched to 14-hour shifts, six days a week. The base hospital treated an endless stream of patients wounded by bombings, rockets, mortar fire, and napalm burns. *"It was an extremely difficult situation to work in,"* he explained. *"Our patients weren't just soldiers. We sometimes had to treat civilians—they bore the brunt of the war."*

One of the most harrowing memories he shared was of a young Vietnamese boy brought into the hospital after playing with an unexploded mortar shell. *"The kid was no more than 7 or 8 years old. He'd lost both legs and an arm. I'll never forget the wailing of his mother. It's a sound you never, ever forget."*

Dr. Kenner paid special tribute to the nurses and medics who worked alongside him. *"There was Maisy O'Laughlin, a nurse who had served in WWII. She was tough as nails but had a heart as big as a canyon. She spent countless hours writing letters home for the soldiers and listening to their stories."* Even Maisy, known for her steely demeanor, broke down when the boy was brought in. *"It got to her—for a couple of seconds, she lost her composure. It was just too much."*

The medics earned Dr. Kenner's deepest respect. *"They were top-notch. They assisted in surgery, removed dead or infected tissue, set broken bones, and changed dressings. Many of the soldiers would not have survived without their skills and dedication. I have the greatest respect for those medics."*

Despite the overwhelming workload, the staff found ways to cope. A company clerk named Ronny Greshner—affectionately known as "Gresh"—became a lifeline for the medical team. *"Gresh could get you just about anything,"* Dr. Kenner said with a smile. *"His persistence and wheeling and dealing helped us get much-needed supplies and save countless lives."*

Holidays were especially difficult. The demands of war left little room for celebration. *"You don't plan on doing a six-hour operation on Christmas Eve because a soldier stepped on a booby-trapped landmine and blew his leg to shreds, but that's what you do,"* he said. *"When you weren't in surgery, you drank. We drank to ease the pain and trauma we saw every day."*

Reflecting on his time in Vietnam, Dr. Kenner acknowledged the emotional toll the experience has taken. *"I thought it made me a better surgeon when I got back home,"* he said. *"But for more than 50 years, I've been haunted by the lives I changed—the limbs I had to amputate. I wonder, was it really for the better? I did what I thought I had to do."*

The horrors of war have left an indelible mark on Dr. Kenner's life, but so too has the profound sense of duty and compassion he demonstrated in the most challenging of circumstances.

Dr. Jim Kenner, thank you for your service to our great country, and welcome home.

Sergeant Dick McCarthy

United States Army (1966-1968)

Sergeant Dick McCarthy's service during the Vietnam War exemplifies the courage and resilience of soldiers who faced constant danger in the field. Serving from 1966 to 1968, McCarthy was thrust into one of the most intense and controversial conflicts in U.S. history. His journey from Fort Dix, where he completed basic training, to the battlefields of Vietnam left an indelible mark on his life.

At just 20 years old, McCarthy landed in Bien Hoa, a critical base in South Vietnam. The war's brutal realities quickly became apparent. When asked about holidays during his deployment, McCarthy's response was stark. ***"Didn't pay that much attention to the holidays. We were out fighting in the field. You were lucky to have something to eat."***

The fear and unpredictability of combat were ever-present. McCarthy shared a chilling story from his time in Vietnam. ***"I was in a jeep with a soldier named James H. Taylor. I was in the back, and he was driving. We were the last vehicle in our brigade. I took out a grenade and lobbed it over my shoulder. It exploded not too far from a Vietnamese man riding his bike. We heard on the radio that the front of the brigade was being attacked. We rounded a corner and came face to face with about ten VC, dressed in their black pajama uniforms with guns. We were as scared to see them as they were us, and we just zoomed by them."***

168

The brutality of the North Vietnamese Army (NVA) and Viet Cong was a constant reminder of the stakes. McCarthy recounted hearing a desperate call over the radio from a soldier surrounded by enemy forces. *"A general was trying to calm him down. Next thing we heard, the voices of the VC that had them surrounded. They killed everyone and skinned a Vietnamese interpreter alive."* The horrific tactics of the enemy weighed heavily on the minds of American soldiers. *"The NVA was brutal. They'd throw guys out of helicopters to make the last guy talk. Sometimes you see things you're not supposed to, and you just don't want to fight anymore."*

The fear of dying without their bodies being recovered was a pervasive concern. *"It's one thing to die, but thinking your body is never going to be recovered weighed heavily on a lot of us,"* McCarthy said.

Despite the fear and trauma, McCarthy's bravery in combat earned him the Silver Star. During a mission, a mine was detonated near his vehicle, giving him a concussion and wounding two of his comrades. The citation from the Department of the Army on February 17, 1967, described his actions:

"Private McCarthy heroically stood up in the line of fire, recovering from the initial shock and with disregard for his own safety, and brought a rapid rate of deadly automatic fire upon the Viet Cong enemy, causing them to flee. Private McCarthy's immediate, timely, and decisive action undoubtedly saved the lives of his comrades and others in the convoy."

Reflecting on his time in combat, McCarthy described the chaos and confusion of battle. *"It's about survival. Sometimes the smoke is so thick, you can't see the enemy. You try to maneuver and not get surrounded. Once you're surrounded, you're [expletive]."*

169

Though the war was filled with grim memories, there were moments of lightness that lifted the spirits of the troops. When asked about entertainment in Vietnam, McCarthy's face brightened as he spoke of Martha Raye. *"Martha Raye... boy, was she something!"* Raye, an actress and singer, dedicated much of her time to performing for the troops. She earned the affectionate nickname "Colonel Maggie" for her work with the Green Berets and her tireless efforts to boost morale. Unlike other entertainers, Raye often went into forward areas, putting herself at risk to reach soldiers on the frontlines. Her visits were a rare and welcome reprieve from the horrors of war.

McCarthy's thoughts on the war and its leadership reflected the frustrations felt by many soldiers. *"You can't have politics in war. Plain and simple. Johnson [President Lyndon B. Johnson] and McNamara [Secretary of Defense Robert McNamara]... you don't run a war with rules,"* he said in disgust.

Returning home from Vietnam was another challenge. Like many veterans of the conflict, McCarthy struggled with the reception from a divided nation. However, his service, courage, and dedication remain a testament to the strength of the soldiers who fought in one of America's most difficult wars.

Sergeant Dick McCarthy, thank you for your service to our great country and welcome home.

Captain Mark McMahon

United States Marine Corps (1968–1973)

Mark McMahon served his country in the US Marine Corps from 1968 to 1973, resigning his commission as a Captain. A native of Whitinsville, Massachusetts, he attended Officer Candidate School (OCS) at Quantico, Virginia, before being assigned to the 2nd Marine Division at Camp Lejeune in Jacksonville, NC. His early service took him on deployments to the Mediterranean, followed by language training at Fort Bragg, NC. In January 1971, he arrived in Vietnam, landing in the dead of night. *"It was so hot and humid, and we had our wool jackets on,"* he recalled. *"We had a battalion commander whom I thought had the mindset to run in battle… made sure I stayed right behind him so that he couldn't."*

One of the most crucial lessons Captain McMahon learned in combat was the necessity of both respecting the enemy and practicing strong leadership. *"The enemy always knew where we were,"* he stated. *"They flowed like water… went where they wanted to… they constantly looked for laxness… when your guard was down… they were trying to kill us, and we were trying to kill them… you had to respect the enemy."*

The North Vietnamese Army (NVA) was a disciplined, resourceful, and highly adaptable fighting force. Their combat strategy relied heavily on guerrilla tactics, utilizing the dense jungles of Vietnam to their advantage. The NVA and Viet Cong used hit-and-run attacks, ambushes, and intricate tunnel systems to move troops and supplies undetected. They understood terrain like no other, blending into the environment and launching attacks with calculated precision. The Ho Chi Minh Trail, an elaborate supply network stretching through Laos and Cambodia, allowed them to reinforce their forces and sustain long-term combat operations despite heavy American bombing campaigns.

The NVA was also patient. They exploited weaknesses in U.S. strategies, particularly the reliance on fixed positions and predictable movement patterns. *"Strong leadership is essential in combat,"* Captain McMahon emphasized. *"You have to bring a sense of fearlessness... anticipate and learn from your mistakes... avoid repetition... don't take the same trail twice... if you establish a pattern, the enemy will find out and you're done."*

To counter the NVA's tactics, U.S. forces relied on overwhelming firepower, air superiority, and rapid mobility. Helicopters played a key role in deploying troops quickly into hot zones and extracting them just as fast.

Captain McMahon recounted a day that could have ended in tragedy. *"We got shot down... a couple of soldiers stepped on booby traps... helicopters wouldn't fly into combat. Protocol is for helicopters to identify the smoke on the ground from their soldiers, which is based on wind direction and location. The pilot lost the smoke... came in at 50 to 100 feet at 11 knots... we went down and landed at the bottom of the skids."* The NVA had anticipated this move and had already positioned themselves beneath the helicopter, preparing for a rapid assault. *"I got out of the chopper and got one injured soldier, then hustled back and got another one... happened so fast... held them off until our gunships finally came."*

Christmas in Vietnam was just another day—if anything, it meant heightened alertness. *"A day like any other—Ho Chi Minh's birthday... we always had to be on guard. Once you're out in the field, your head's on a swivel. If you were outside the wire, you had to be extremely mindful of what's going on... when you're in the field with the Viet Cong enemy, that's the life you live. When you're in and out of engagement... when you hear that metallic sound, you need to be prepared to do something. Even when I walk in and out of doors now, I get that feeling."*

When asked about a mentor who had an impact on him, Captain McMahon didn't hesitate: *"Major John Raymond, our deputy regional advisor. Set a great example... unbelievably smart... understood that I didn't know much... wouldn't ask anyone to do anything he didn't do himself."*

Was he ever afraid? *"Oh sure, all the time... they came at us at night... we fired illumination rockets into an area... always relieved to see dawn breaking."*

Reflecting on his return home, he said, *"Nobody spit at me... people tended to treat you like you were in a crypt... just got on with my life."*

His time in service shaped his approach to life. *"It set the pattern for my life... I don't do anything half-assed. I'm always seeking a challenge. Climbed Mount Kilimanjaro over 20 years ago... ran 10 marathons... you either lead, follow, or get out of the way!"*

For his valor, Captain McMahon was awarded two Vietnamese Crosses of Gallantry.

Captain Mark McMahon, thank you for your service to our great country and welcome home.

Sergeant Major John Mennitto

U.S. Army (1965-1989)

It was a distinct privilege to spend time with this veteran, Sergeant Major John Mennitto. Even after years of service, he still carries himself with the presence and discipline of a soldier. His demeanor commands immediate respect, a testament to his 24 years of distinguished service in the U.S. Army.

Growing up in Somerville, Massachusetts, in a military household, John Mennitto was no stranger to discipline and service. His father, a Navy veteran of World War II, instilled in him a strong sense of duty from a young age. *"I learned discipline early,"* he recalled, *"marching in parades and participating in the Sons of The American Legion."* At an event, he met a WWII veteran who shared stories of his military service, sparking a curiosity that would shape John's future. On December 7, 1965, at just 17 years old, John enlisted in the U.S. Army.

John began his military journey at Fort Dix, New Jersey, for basic training. His first assignment was at Fort Lewis, Washington, where he served as a company clerk with the 4th Infantry Division. Despite his initial administrative role, John's leadership potential quickly became evident. By the age of 19, he had earned the rank of Drill Sergeant.

Reflecting on those early days, John remembered his first sergeant vividly. *"He was a tough guy,"* John said, *"and he had a Merrill's Marauders patch on his arm."* Merrill's Marauders were a legendary U.S. Army unit specializing in jungle warfare during World War II. Inspired by such role models, John pushed himself

174

through Drill Corporal School, an intense training program he described as *"basic training on steroids."*

His career soon took off. He was deployed to Thailand and other countries on highly classified missions, the details of which remain confidential to this day. Within just four years, he was promoted to Staff Sergeant. In early 1972, he received orders to deploy to Vietnam.

When asked about spending the holidays away from home, John said, *"It was tough. But the good thing was, we were all in the same boat."* One Christmas in Vietnam stood out in his memory. Stationed in Eastern Long Binh, his unit was supposed to enjoy a ceasefire during the holiday season. However, the Viet Cong had different plans. *"They crawled under our wire and used a satchel charge to blow up helicopters and fuel tanks,"* he recounted. *"Our choppers corralled about 40 VC in a place we called Monkey Valley and let loose on them."*

Another vivid memory took place in Can Tho. John described a chilling incident where the Viet Cong hijacked an American ambulance, killing everyone inside. Disguised as medics, they breached the base's barriers and began throwing grenades from the ambulance. *"We found out just in time,"* John said. *"I grabbed my M-72 LAW rocket launcher and fired. They were no more."*

Throughout his 24-year career, John excelled in various roles, including as a military recruiter. At one point, he was responsible for seven recruiting stations. His assignments took him to prestigious institutions such as West Point and the Army War College. In Hawaii, he served as a First Sergeant in the 25th Infantry Division before being promoted to Sergeant Major and attending the U.S. Army Sergeant Major Academy at Fort Bliss, Texas.

John's leadership journey continued at Fort Ord, California, with the 7th Infantry Division. There, he served as the G-1/A6 Sergeant Major, overseeing critical operations. When asked about the importance of mentorship, John credited much of his success to his mentor, Ted Jackson. *"Without mentors, I wouldn't have been successful,"* he said. *"Ted Jackson was a great one. He still is."*

John's return from Vietnam was a bittersweet experience. Flying into San Francisco in uniform, he encountered hostility from civilians opposed to the war. *"We had to take a short helicopter ride to Oakland,"* he recalled. *"The people onboard started glaring at me and making nasty comments."* Just as the tension escalated, two unexpected allies stepped in, members of the Hell's Angels motorcycle gang. *"They yelled, 'Enough! Now you people apologize to the sergeant.'"* John said. *"It was unbelievable. I'll never forget that."*

Looking back on his service, John expressed deep gratitude for his time in the Army. *"I wouldn't have done it any other way,"* he said. *"If I could have stayed in the Army until I died, I would have."*

Among his many accolades, two stand out as his most cherished: the Army Ring, awarded by General Maxwell Thurmond in 1980, and the Sergeant Major's Ring, which he received in 1987. These honors symbolize his exceptional service and dedication.

Sergeant Major John Mennitto, your unwavering commitment to our nation and your courageous service inspire us all. Thank you for your service to our great country.

Welcome Home.

Etta Moran

American Red Cross Donut Dollie (1967–1970)

Etta Moran served her country with extraordinary bravery, compassion, and dedication as a Red Cross Donut Dollie during the Vietnam War. From 1967 to 1970, she brought hope and comfort to U.S. service members overseas. Donut Dollies were American Red Cross workers whose mission was to boost morale, and Etta embodied this spirit with grace and courage.

Etta was born and raised in Galena, Illinois, a picturesque town in the northwest corner of the state. Known for its historic charm and rolling hills, Galena was a tight-knit community where neighbors knew one another by name. Galena's history as the home of Ulysses S. Grant, the Civil War general and later president, may have also sparked Etta's sense of duty to serve her country.

After graduating high school, Etta attended the University of Chicago, where she received both her undergraduate and Master's degree at one of the most prestigious institutions in the country. In the mid-1960s, it was uncommon for women from small towns to pursue higher education, particularly at elite universities. Women at the University of Chicago faced unique challenges in an era when gender roles were still rigidly defined.

Etta's journey with the Red Cross began after completing her graduate studies. Meeting the rigorous standards to become a Donut Dollie, she was deployed to Vietnam in the summer of 1967. Her initial excitement turned to apprehension upon arrival. *"The heat and the stench when we landed in South Vietnam were overwhelming,"* she remembered. *"And the bus they picked us up*

177

in had barbed wire on the windows to stop grenades. That's when I thought, 'What have I gotten myself into?'"

Her first mission was a trip to a firebase by helicopter. *"I was scared to death,"* she said. *"We flew low, and I just prayed we wouldn't get hit."* Upon landing, she and her fellow workers immediately went to work comforting soldiers, handing out donuts, and bringing smiles to the faces of men who hadn't seen home in months. *"We had those soldiers grinning from ear to ear,"* she said proudly.

One of the most emotional moments of Etta's service came during a visit to a hospital ward in Phu Loi. *"I walked in and my heart sank... my brother Billy was lying in a bed, shot in the chest,"* she recounted. Overcome with shock, she fled the room, gasping audibly and catching the attention of everyone around her. The head nurse followed her outside, angry at first, but softened when Etta explained, *"That's my baby brother Billy in there, shot in the chest."* The nurse embraced her and reassured her, saying, *"He's going to be fine. Now get back in there and do your job."* That moment, filled with both heartbreak and determination, defined the personal and emotional sacrifices that Red Cross workers often endured.

Etta spent holidays in Vietnam alongside the troops, finding purpose in bringing joy to young men far from their families. *"These were just boys, like my brothers,"* she said. Her time as a Donut Dollie was both rewarding and heartbreaking. *"It was an honor to serve,"* she reflected. *"We saw a lot that wasn't pretty, but we got through it. My heart still aches for those young boys we lost."*

At 76 years old, Etta lives in Mashpee, Massachusetts with her husband Jim, her partner of 53 years. Etta Moran, thank you for your service and for bringing hope to countless soldiers during one of the most challenging periods in our nation's history.

Sergeant Carl Rosetti

United States Army (1967–1969)

Sergeant Carl Rosetti served his country with courage and resilience as a member of the United States Army during the Vietnam War. A native of Brooklyn, New York, Carl grew up in a tough neighborhood, excelling in sports and dreaming of playing professional baseball for the New York Yankees. His life took a dramatic turn the day he was drafted. *"I went to Vietnam because I didn't really have a choice. My grandfather fought in WWI, and my father was a WWII veteran. It was my turn,"* he explained.

After completing basic training, Sergeant Rosetti was assigned to the Special Forces Unit – Studies & Observation Group, a highly classified and demanding role. Reflecting on his experiences, he described his unit as a mix of *"major league, sick people"* whose missions were often too extreme to recount in detail.

The jungles of Vietnam presented constant challenges, beyond the ever-present threat of the Viet Cong. *"We were in the jungle for three to four weeks at a time. Poisonous snakes, giant centipedes, and even tigers were everywhere,"* he recalled. On one mission, their unit ventured so far behind enemy lines that their own helicopter mistook them for the enemy and fired at them. *"We didn't think we could go that far without getting killed by the VC."*

The fear of capture by the Viet Cong loomed heavily over the troops. *"We'd heard about a helicopter pilot they shot down. They beat him mercilessly, tortured him with a hot poker, and threw him into a box with rats, hands and feet tied. That was always on our minds,"* he said grimly.

179

Sergeant Rosetti also spoke about the physical and psychological toll of missions. One haunting memory involved an American soldier stuck in jungle mud up to his neck. *"He was a tunnel rat and got caught in a torrential rainstorm. The mud just sucked him down. He stayed there all night, with VC patrols walking past him, snakes slithering by, and mosquitoes biting his face until it was swollen. When we found him, he was in bad shape."*

The intense violence and relentless nature of their operations left little room for moments of levity or reflection. When asked about good memories, Sergeant Rosetti responded with a mix of frustration and pain. *"Good memories? Jesus Christ, we killed people. It wasn't a country club. Every day was about receiving orders, searching, destroying, and killing—or being killed. You became like a robot. Nothing else mattered except completing the mission."*

He also debunked the sanitized narratives sometimes presented in documentaries or public accounts of the war. *"I laughed my ass off when I watched a documentary about the Ho Chi Minh Trail. A U.S. general claimed he visited troops in the jungle during the holidays. Pure BS. They couldn't even find us, let alone visit."*

Sergeant Rosetti's time in Vietnam continues to affect him. He candidly shared that he suffers from severe PTSD, often waking in cold sweats from nightmares. Despite these struggles, he has found stability in his life with his wife of 35 years, Mary, in Bourne, Massachusetts.

Sergeant Carl Rosetti, thank you for your service to our country and welcome home. Your sacrifices and those of your comrades will never be forgotten.

Lieutenant John Shoemaker

U.S. Army (1968-1972)

It was an absolute honor to interview Lieutenant John Shoemaker for Veteran Spotlight. Born in Fitchburg, Massachusetts, Lt. Shoemaker attended the University of Massachusetts (UMass), where he earned a Bachelor of Business Studies (BBS) in Management before enlisting in the Army in 1968. He went on to serve his country for four years as a Combat Platoon Leader and Company Commander in the 196th Light Infantry Brigade of the Americal Division, stationed in Chu Lai, Vietnam.

Lt. Shoemaker completed basic training at Fort Dix, New Jersey, followed by Advanced Infantry Training at Fort Polk, Louisiana. He then underwent Airborne and Officer Candidate School training at Fort Benning, Georgia, before advancing to jungle warfare training in the Panama Canal Zone.

When Lt. Shoemaker deployed to Vietnam, his first-born son was only three months old. Upon arrival, he underwent two weeks of in-country training before being assigned to a firebase on Hawk Hill and subsequently to Hill 251, located just west of Chu Lai. His first day on patrol left an indelible mark.

Recalling that day, he shared, *"My commanding officer told me, 'You're just an observing officer.' We were marching through a rice paddy, and I was fifth in line. The radio operator was in front of me. I was mesmerized as I waded through dark brown, putrid, god-awful-smelling water filled with water buffalo, human excrement, and leeches. I thought, 'I'm not watching*

181

Walter Cronkite anymore.' Then, it seemed like the whole world blew up."

A soldier ahead of him stepped on a booby trap, causing a massive explosion. The blast severed one man's feet and nearly split the lieutenant in two. *"Suddenly, I was in charge,"* Lt. Shoemaker said. *"All our guys had zero combat experience. One of the soldiers came up to me and said, 'Okay, Lieutenant, what do we do now?' I'd only been on patrol for two hours, and I thought, 'How am I going to last a year?'"*

Determined to keep his men alive, Lt. Shoemaker resolved to take the fight to the enemy. *"I was scared every day,"* he admitted. *"But I told myself, 'I'm taking the battle to the enemy.' We were in the jungle for three to four weeks at a time. I was always afraid of losing a limb or being captured. I resigned myself to never being taken alive."*

When asked about holidays during his deployment, Lt. Shoemaker recalled one memorable experience: seeing Bob Hope perform. *"There were no holidays for me,"* he said. *"But I did have the chance to see Bob Hope. It was a life experience. I even got to meet him years later."* He shared that he and his men from Hawk Hill traveled a long distance to attend the show, but the journey was fraught with tension. *"All I could think about was, 'To hell with Bob Hope—what happens if we get ambushed?' We couldn't carry guns; only the MPs escorting us were armed. It would have been the perfect time for an attack."*

Lt. Shoemaker reflected on the pain of losing soldiers under his command. *"It's simply devastating,"* he said. *"You constantly think, 'What could I have done? What should I have done?' It eats at you."* He lost five soldiers—two to gunfire and three to booby traps.

"One of the losses that still haunts me was a young soldier whose parents were both doctors. He was in medical school studying to become a doctor himself. His parents encouraged him to enlist, saying he should do something meaningful for his country. During a firefight, he took a bullet to a major organ. He died from shock in the helicopter. An incredible human asset was tragically lost."

When asked about his thoughts on the Vietnam War protesters, Lt. Shoemaker said, *"I rejected it at the time because it showed our country in a bad light and caused division. It destroyed our country's morale, unity, and patriotism."*

Lt. Shoemaker shared his reflections on military service: *"I felt it was my duty to serve. My father was in the Navy during World War II, and he inspired me. Being a Combat Platoon Leader in Vietnam defined who I am. It changed who I was and shaped who I would become. I would do it all over again the exact same way."*

Lt. Shoemaker's bravery and dedication were recognized through numerous awards: three Bronze Stars, two with Valor; the Air Medal; two Army Commendation Medals, one for Valor; and the Purple Heart.

Thank you, Lieutenant John Shoemaker, for your selfless service to our great country.

Welcome Home.

Sergeant Don Silvia

United States Army (1970 – 1971)

Sergeant Don Silvia served his country with bravery in the United States Army during the Vietnam War. A member of the 23rd Infantry Division, also known as the American Division, he was assigned to B-Battery and D-Battery, 6th Battalion, 11th Artillery, and saw heavy combat throughout his tour.

A graduate of Lawrence High School in 1968, Sgt. Silvia began his military journey at Fort Dix, NJ for basic training before continuing to Advanced Infantry Training at Fort Sill, OK. His distinction as the first Massachusetts draftee for the 1970 draft made headlines. *"I was a nervous wreck when they called me up to the front of the line…all the big Boston news stations were there—Channel 5, Channel 7—all of them,"* he recalled. Despite the attention, he excelled in training, earning the 4th highest rating out of 200 recruits.

Sergeant Silvia's deployment brought him to Da Nang, Vietnam. His first assignment was to San Juan Hill, about 1,300 feet above sea level and just 12 miles from Duc Pho. This strategic hill, overlooking the South China Sea, absorbed relentless enemy fire from the Viet Cong. *"We had six guns on the hill with 60 guys—102 Howitzers. You could fire in 360 degrees. The infantry was down at the lower level of the hill, and every night we rotated guard duty every two hours. We had barbed wire, spotlights, and barrels of kerosene to hold off an enemy advancement,"* he shared.

The dangers were ever-present. *"Of course, I was afraid! You could see the incoming mortar rounds coming. The Viet Cong*

184

shot from the villages, but we couldn't fire back because they had civilians."

Sgt. Silvia experienced firsthand the hardships of monsoon season. *"We were a Jump Battery, supporting the infantry. It was monsoon season, and we had to abandon our guns and position in the field. The VC knew there would be flooding. We followed the ARVN (Army Republic of Vietnam) up the mountains. It rained all night long. We had tents, but nobody slept. The next day we went back, and all our guns were sunk in the mud."*

He vividly remembered the nights being the most terrifying. *"The VC always came out at night. I slept in a bunker four feet underground. It was hot too. Our guns could shoot seven miles away, but even a couple of millimeters off could mean killing someone by mistake."*

San Juan Hill in Vietnam was a key strategic location during the conflict. It was heavily contested due to its elevation and proximity to supply routes. U.S. forces on San Juan Hill faced relentless mortar attacks and sniper fire, making it a precarious post to defend. Despite the hill's challenges, Sgt. Silvia and his fellow soldiers maintained their positions to provide critical artillery support to infantry units operating in the surrounding areas.

One of the most lasting and devastating legacies of the Vietnam War was the use of Agent Orange, a herbicide and defoliant used by U.S. forces to clear dense jungle cover. Unfortunately, its toxic effects have led to severe health issues for many veterans, including Sgt. Silvia. *"My lungs are still bad from all the Agent Orange we dropped in the mountains,"* he shared. The chemical exposure caused long-term respiratory problems and other ailments that continue to affect veterans to this day.

Returning home from Vietnam was a moment Sgt. Silvia will never forget. *"When you get on that plane, everyone is*

185

quiet...when we got airborne, everyone went crazy with applause," he recalled. His service was recognized with numerous awards, including the Bronze Star, National Defense Service Medal, two Army Commendation Service Medals, and the Vietnam Service Medal.

Following his military service, Sgt. Silvia continued to serve his community as a firefighter and EMT for over 30 years. He remains grateful for the bonds formed during his time in the Army. *"Holidays were just like any other day. You make a lot of good friends, and you share it with them. I wrote back and forth to my parents and got lots of care packages,"* he said.

One of his most significant influences during his service was Sgt. Jim Ashwell, whom he described as a mentor and a friend. *"He was just a good guy who took me under his wing. He even came to Falmouth to visit me after Vietnam."*

Reflecting on his service, Sgt. Silvia expressed pride in his contributions. *"It was a great honor to serve my country. It was worthwhile going and helping other people. Much different from today."*

Thank you, Sergeant Don Silvia, for your bravery and unwavering commitment to your country. Welcome home.

Specialist 5 Dave Smachetti

United States Army (1965-1967)

Specialist 5 Dave Smachetti served his country with distinction during the Vietnam era, working as an aviation mechanic responsible for maintaining critical flight equipment in military aircraft. His service took him across the Atlantic to Manheim, Germany, which became a vital location for U.S. military operations during the Cold War. His journey from a small-town boy in Adams, Massachusetts, to a soldier entrusted with the safety of military flights highlights his dedication, resilience, and adaptability.

Smachetti began his military career with basic training at Fort Dix, New Jersey, an experience he described as *"baffling, confusing, and overwhelming"* for a young man from Adams. But his perseverance paid off, earning him an Expert Marksmanship Commendation before he continued his training as an aviation mechanic at Fort Gordon, Georgia. Aviation mechanics played a critical role in military operations, ensuring that flight controls, homing devices, and other essential systems were in top condition. *"In your job, you simply couldn't be careless. You had to be on top of your game,"* Smachetti explained.

His first major assignment took him to Manheim, Germany, home to one of the largest U.S. Army installations in Europe at the time. During the Cold War, the U.S. military presence in Germany was critical to countering the Soviet threat. The bases in Germany, including those in Manheim, served as logistical hubs, training grounds, and forward operating locations for troops ready to respond to any potential conflict in Europe.

187

Smachetti's journey across the Atlantic to Germany was no small feat. Traveling aboard the USNS Darby with 3,400 other troops, he endured the harsh winter seas during Christmas and New Year's. *"It was kind of a sad time being away from home,"* he recalled. The camaraderie he found among his fellow soldiers became essential to his daily life and his ability to cope with being so far from home. *"We had our regular day jobs, but we hung around a lot together. That made things a lot easier."*

Life in Manheim wasn't without its challenges. Security at U.S. bases during the Cold War was incredibly tight, especially as tensions between the United States and the Soviet Union were at their height. Smachetti recalled the readiness drills and the constant state of alertness required of those stationed there. *"The Vietnam War was going on, so when that red light went on at night, you had to be ready to go. You had a certain, special code to get to the base, and security was air-tight."*

The role of aviation mechanics like Smachetti was critical to ensuring that military aircraft were mission-ready at all times. Whether it was routine maintenance or urgent repairs, the work of mechanics ensured the safety of pilots and the success of missions. Any mistake could be catastrophic, which is why the job demanded precision and focus. *"Knowing I had guys I could depend on really helped a great deal,"* he said.

Smachetti shared a memorable story from his time crossing the Atlantic during a winter storm. *"We had waves, wind, cold— everything you could imagine. You had to be on your watch station no matter what. At the end of one watch, the saltwater had actually eroded the stitching off my boots. We put up with all kinds of weather in the Atlantic,"* he recalled.

Despite the hardships, Smachetti expressed deep pride in his service. *"I feel extremely honored. I feel great. Very proud of my service,"* he said. His pride is reflected in his continued dedication to Veteran organizations. He has been a member of the Adams American Legion Post 160 for an astounding 52 years and was recently honored with the prestigious John Quigley Award at the American Legion Regional Convention in Marlboro, Massachusetts. Upon completing his service, he was also awarded the National Defense Service Medal.

Specialist 5 Dave Smachetti's contributions as an aviation mechanic ensured the success and safety of countless military operations during a pivotal era in world history. His experiences reflect the sacrifices and dedication of those who served during the Cold War and the Vietnam era, standing ready to defend freedom at any moment.

Specialist 5 Dave Smachetti, welcome home, and thank you for your service to our great country.

Corporal Walt Stillman

United States Army (1968-1970)

Corporal Walt Stillman's service in the United States Army during the Vietnam War was marked by both unimaginable horrors and moments of humanity that brought light to a dark time. Drafted right out of high school, Stillman's path to military service was one of escape from a troubled home life. Born in Mobile, Alabama, he recalled, *"I was 100% sure I was going anywhere, whether it be Vietnam or Timbuktu. I just needed to get out."*

After completing basic training at Fort Lewis, Washington, Stillman was sent to Vietnam in 1968 and assigned to the 8th Field Hospital, first in Nha Trang and later in An Khe. These hospitals played a critical role in treating the overwhelming number of wounded soldiers brought in from the frontlines. The field hospitals were often the first stop for severely injured troops before they were evacuated to larger medical facilities or sent home.

At the hospital, Stillman served as a company clerk, a role that required him to handle everything from paperwork to phone calls to requisition orders. *"It was a 5-pronged job. Everything came through me,"* he explained. However, the job quickly evolved beyond administrative duties due to the constant flow of casualties. *"Because we were so short-staffed, I'd have to help out in the operating room, retrieving blood, new bandages, medicine. Sometimes there was so much blood on the floor that you stepped in it, and it scattered just like stepping into a puddle."*

The trauma of witnessing the aftermath of combat left a lasting impact on Stillman. He shared a haunting memory from his second day on the job. *"I had to wheel out a body of a Marine that didn't make it. I got to the tent we used as a morgue and pulled the sheet off enough to see that Marine's face. I had never seen a*

190

dead person before. I just froze for about 20 seconds. He was just a kid—God, it hit me like nothing I've ever experienced. He was never going to go home to his parents, start a family, have a birthday, or celebrate Christmas. I still can't get that face out of my mind."

Field hospitals like the 8th in Nha Trang and An Khe were overwhelmed with casualties during the height of the war. The medical staff, including doctors, nurses, and clerks worked tirelessly to save lives under incredibly difficult conditions. The sound of screams and cries from the operating rooms became an inescapable part of daily life. *"You'll hear the screams and cries of the operating room until the day you die. It's haunting,"* Stillman recalled. The emotional toll was heavy, and many staff members turned to alcohol to cope. *"When I first got there, I never understood why the doctors drank so much. Then I realized why after I'd been there a month."*

Amidst the trauma, one bright spot stood out—the visit from Bob Hope. The legendary entertainer made it his mission to boost the morale of American troops during the holidays, performing in USO shows and visiting hospitals to lift spirits. *"My goodness, what presence that man had,"* Stillman said with admiration. *"He put everybody at ease. He was tender, kind, and funny when he spoke to the injured. He had a joke for everything."*

Hope's ability to bring laughter to soldiers facing the horrors of war was unparalleled. His visits were a reminder of home, a brief escape from the reality of combat. Stillman vividly remembered one interaction. *"I heard him whisper to a nurse and wink, 'That's what they want, right sis?' The morale boost he gave all of us—you just can't put it in words."*

Returning home from Vietnam was another challenge for Stillman. Like many Vietnam veterans, he faced hostility from anti-war protestors. *"My family and friends prepared me prior to my coming home,"* he said. However, he wasn't completely spared from negative encounters. ***"I had an incident when I came back to the airport in San Diego. But I told a few to [expletive] off in a really loud voice, and they backed down. After all I went through, I never gave them the time of day or much thought."***

Reflecting on his service, Stillman acknowledged the deep and lasting impact of his time in Vietnam. His experiences at the 8th Field Hospitals in Nha Trang and An Khe represent the harsh realities faced by medical staff during the war, who worked around the clock to save lives despite overwhelming odds.

Corporal Walt Stillman, thank you for your service to our great country and welcome home.

1st Lieutenant Brian Sullivan

United States Army, 93rd Military Police Battalion (1970–1971)

Brian Sullivan never fought in the jungles of Vietnam, but for nine months, he experienced the brutal realities of war as a military police officer in the chaotic streets and back alleys of Qui Nhon. Serving as a 1st Lieutenant in the Army's 93rd Military Police Battalion from October 1970 to July 1971, his mission was to maintain order in a city overrun by corruption, black market activity, and violence. *"It was an absolute cesspool, and we did what we had to maintain law and order,"* he recalled.

A native of Hingham, Massachusetts, Brian grew up in a military family and went through ROTC at Northeastern University. After earning a Master's degree from Boston State Teachers College in 1969, he advanced to officer training and was soon deployed to Vietnam. There, he led a platoon of MPs, often teaming with Korean and Vietnamese police to patrol the city's volatile streets.

Qui Nhon was a supply port and a hub for criminal activity, and as American troops began to withdraw, the situation grew even more dangerous. *"The more troops that left, the more dangerous Qui Nhon became. Crime and corruption were everywhere. The city was ready to explode at any minute,"* Brian said. His leadership extended to coordinating the use of V-100 armored personnel carriers and riverboats to protect the harbor.

Respect was critical in an environment as hostile as Qui Nhon, and Brian earned it through a tough, no-nonsense approach. *"I was bad, and I knew I was bad. When you walk into a bar full of drunk people, you had to create an image. The troops respected me in the alley—that's where respect mattered most,"* he explained. But the toll of such constant tension wasn't lost on him. *"I lost my way many times, lost my focus, but God reached down and pulled me out when I needed it most."*

Brian's role as an MP also involved difficult moments, including escorting U.S. soldiers convicted of serious crimes. One memory that stuck with him was from July 1971, when he led a guard detail transporting 10 prisoners from Long Binh Jail to San Francisco. *"At the airport, my guards and I were insulted and derided by the flower children, who empathized with the prisoners without knowing their crimes. The welcome home we received that day will stick in my mind forever—it sucked,"* he said.

Despite the challenges of his service, Brian deeply admired his wife, Betsy, for her strength and sacrifices during his time in Vietnam. *"In 1971, Betsy was home with our four kids. When a green sedan pulled up in front of the house, she nearly fainted, thinking it was to notify her of my death. She didn't want to hear anything about the war, but the reminders were constant. Delayed mail drove her to worry endlessly,"* he recalled. *"Everyone focuses on the veteran, but the spouse and family serve, too. Their sacrifice needs to be remembered."*

One of the most powerful moments of Brian's post-war life came during a 4th of July parade. *"We were marching with our Huey Helicopter Float, and I heard the cheers from the crowd. Tears welled up in my eyes. To finally receive the 'Welcome Home' we were deprived of in '71 was overwhelming. Seeing elderly WWII and Korean War vets pushing themselves out of their chairs to stand and salute us is an honor I will never forget,"* he shared.

After Vietnam, Brian taught for a year before continuing his service in the National Guard, earning the Guard's Medal of Valor.

He left the Guard in 1988 and joined the Army Reserves, retiring as a provost marshal in 1993. Survivor's guilt still leads him to the VA for PTSD counseling, but his pride in his service remains steadfast. ***"I am proud to be a Vietnam Veteran and proud of the men I served with."***

Lieutenant Brian Sullivan, thank you for your service and sacrifice. Your dedication to our country and your fellow veterans is deeply appreciated and will never be forgotten.

Welcome Home.

Staff Sergeant Kris Tebbetts

United States Air Force (1965-1968)

Kris Tebbetts served his country in the United States Air Force as a Staff Sergeant from 1965 to 1968. He began his military journey with basic training at Lackland Air Force Base in San Antonio, Texas. Reflecting on that time, he shared, *"I was 17 years old... basic, it was OK... didn't mind it... graduated from high school in June and was like, 'what do I do now?'... went to Florida on vacation, then came home and saw the recruiter."*

His first assignment after basic training was at Keesler Air Force Base in Biloxi, Mississippi, where he learned Morse Code, a critical skill for military communications during the Vietnam War. Morse Code was a vital tool used to intercept and relay intelligence during combat operations. Communicators, often referred to as "Ditty Boppers," played a key role in the war effort by intercepting enemy transmissions and ensuring that critical messages were accurately delivered. Kris Tebbetts embraced this role with dedication, humorously recalling his nickname, *"They called them Ditty Boppers,"* he said with a laugh.

Following his time at Keesler, Tebbetts was selected for Air Force Security and underwent further training at Goodfellow Air Force Base in Texas, where he spent four months learning electronic communications. His expertise in communication technologies was essential in gathering intelligence to support military operations. This training prepared him for his assignment at Clark Air Force Base in the Philippines, where he worked in intercept

196

communications. *"Everything was classified, top secret... we didn't exist," he recalled. His work was part of a covert mission to monitor enemy communications, ensuring U.S. forces had a strategic advantage. Despite the intense secrecy of his mission, Tebbetts fondly remembered the sense of community. "One of the great things that we did was to help build a local school in the community for the kids,"* he shared.

Entertainment overseas played a significant role in maintaining morale. Tebbetts recalled seeing the Everly Brothers perform for the troops, saying, *"They put on a good show."* When asked about holidays abroad, he reflected on making the best of their situation. *"You always miss home... we made them fun, though. I asked my sergeant if we could have a Christmas tree... he said yes, and I went out and bought a six-foot tree... he couldn't believe it... thought I was going to get one about a foot high... we had everyone over... it was fun,"* he reminisced.

After his service in the Philippines, Tebbetts returned to Goodfellow Air Force Base before being assigned to Wakkanai Air Force Base in Japan. Located at the northernmost tip of Japan, Wakkanai was a strategic listening post during the Cold War. *"Loved it there... the people were really nice... we were there listening to communications from Russia... bought a motorcycle and rode my bike all over,"* he said.

However, his service also exposed him to the harsh realities of war. *"When I was at Clark Air Force Base, our buildings were very close to the road, and we saw everything that came and went... never forget the abundance of wounded soldiers that would come by in trucks and ambulances... then you go into the hospitals and see all the amputees... it's something I always think about,"* he shared solemnly.

Reflecting on the anti-war protests of the era, Tebbetts expressed a nuanced view. *"They were right... when I came out, I went to anti-Vietnam rallies... I don't think we should have been there."* Despite this, he took pride in his service. *"The whole thing was a positive experience for me... I was very proud to serve my country, though it was hard coming back when you had all the resentment aimed at you."*

Staff Sergeant Kris Tebbetts, thank you for your service to our great country, and welcome home.

Combat Medic Billy Ray Thomas

United States Army (1968–1969)

Billy Ray Thomas served with unparalleled bravery as a combat medic in the Vietnam War, attached to the 4th Infantry Division. Drafted out of high school in Topeka, Kansas, Billy completed basic training at Fort Leonard Wood, Missouri, and was deployed to Vietnam in October 1968. From the moment he arrived, the challenges of his role were laid bare: medics were told their life expectancy in the field was just 14 days. *"It was seven,"* Billy remembered grimly.

His first day on the job was a baptism by fire. *"We were ambushed by about 100 VC (Viet Cong). My captain was standing right next to me, telling me to keep my damn head down, when he took one in the helmet—gone, just like that,"* he recalled. In the chaos, Billy had to assume leadership, making split-second decisions about who to treat and who needed immediate evacuation. *"Bullets were flying everywhere. I was scared to death, but you didn't have time to think about fear. Decisions had to be made, and lives were on the line. Anybody who says they weren't afraid is either a lunatic or a liar."*

As a combat medic, Billy faced the dual responsibility of providing life-saving care while protecting himself in the midst of firefights. *"Every patrol was a gamble,"* he explained. *"You never knew when or where the next ambush would come. You had to be constantly on guard."* Despite the dangers, he carried out his duties with unwavering dedication, tending to soldiers under fire and often exposing himself to enemy bullets to reach the wounded.

199

Billy described the experience of hearing desperate cries for help during combat as one of the most haunting aspects of his role. *"I'll never forget those shrieking cries of 'Medic! Medic!' Some of the guys I couldn't get to in time—those are the ones that stay with you. Nights are a struggle; it never leaves you,"* he said, his voice heavy with emotion.

The isolation of life as a combat medic added another layer of difficulty. *"We had no entertainment, no downtime. Patrols in the morning and at night, back to camp during the day—if you could even call it a camp. At night, nobody moved. Moving in the jungle at night was a death sentence,"* he explained. Holidays were no better. "Torture," was how Billy described them. *"You couldn't think about home or family because it just made things worse. You focused on the job—anything else could break you."*

One memory stood out vividly: the visit of a chaplain after a deadly firefight. *"We'd lost some men, and the chaplain came to bless the bodies. He paused for a moment over one boy, visibly emotional. It turned out to be his son,"* Billy said, his voice cracking. *"Jesus, I'll never forget that."*

Through it all, Billy maintained his focus on saving lives. *"You didn't have the luxury of breaking down. You kept moving, kept treating the next guy. There were no breaks, no second chances. The 'ground pounders' trusted us, and we had to come through for them."*

For his service, Billy Ray Thomas was awarded the Bronze Star and Purple Heart, but the emotional toll was immense. *"When I came home, I didn't talk about Vietnam with anyone I served with—not once,"* he said quietly. Despite the hardships, he expressed immense pride in his role. *"I did the best I could for my guys. That's what mattered."*

Today, Billy lives in Mashpee, Massachusetts with his wife, surrounded by the love of his three grandchildren. Though the memories of Vietnam remain, he finds solace in family and the gratitude of those who recognize his service.

Mr. Billy Ray Thomas, thank you for your extraordinary courage, sacrifice, and service to our great country. Your story will never be forgotten.

Welcome Home.

Colonel James L. Tow

United States Army (1952-1980)

Colonel James L. Tow's life has been steeped in military service, a legacy passed down from his father, Colonel William M. Tow, who served in the United States Army for 33 years. Born on an Army post at Fort Eustis, Virginia, in 1930, Colonel Tow's military career spanned 28 years, with service in Germany, Korea, and Vietnam. He commanded a Combat Aviation Battalion with over 100 helicopters, playing a pivotal role in one of the most intense conflicts of the 20th century—the Vietnam War.

Growing up in a military family meant that discipline, respect for tradition, and a deep sense of duty were instilled in Colonel Tow from a young age. His childhood years at Fort William McKinley in the Philippines from 1937 to 1941 left lasting impressions. *"It was safe and secure,"* he recalled, describing the daily routine of the flag lowering ceremony, during which everyone on post would stop, disembark from their vehicles, and face the colors. This early exposure to military life shaped his character and future leadership style.

After graduating from the United States Military Academy at West Point in 1952, Colonel Tow embarked on a distinguished career in the U.S. Army. His first assignment took him to Nuremberg, Germany, where he served as a platoon leader in a Mechanized Infantry Battalion. The historic significance of the city was not lost on him—he vividly remembered seeing the stadium where Adolf Hitler held his infamous rallies and the barracks once occupied by the SS Division. Despite the somber history, he recalled the warmth of German culture during the holidays. *"The holidays in Germany were absolutely wonderful with all the culture, music, and tradition,"* he said.

202

Following his time in Germany, Colonel Tow pursued flight school at Gary Air Force Base in Texas, where he learned to fly a Cessna L-19. His first flight assignment was at Fort Meade, Maryland, where he supported air defense operations. His career in aviation would eventually lead him to Vietnam, where he commanded a Combat Aviation Battalion during some of the most critical moments of the war.

Colonel Tow's time in Korea, stationed at Camp Kaiser north of the 38th Parallel, was marked by the tense reality of the Cold War. *"We were there, and we were prepared,"* he said. The post-war environment in Korea required constant vigilance, as tensions between North and South Korea remained high.

However, it was his service in Vietnam that left the deepest impact. *"Vietnam was a totally different deal,"* Colonel Tow said. His battalion played a crucial role in providing air support, transporting troops, evacuating wounded soldiers, and conducting reconnaissance missions. The use of helicopters revolutionized modern warfare during the Vietnam War, giving the U.S. military a mobility advantage in difficult terrain. However, flying missions in hostile territory came with immense risks.

One of Colonel Tow's most haunting memories from Vietnam occurred during the Tet Offensive—a major coordinated attack by the Viet Cong and North Vietnamese forces in early 1968. *"During the Tet Offensive at Christmas time, I was in my helicopter 5,000 feet above the ground, looking at the burning villages as a result of Viet Cong brutality. It was something I would never forget,"* he said. The Tet Offensive marked a turning point in the war, exposing the vulnerability of U.S. and South Vietnamese forces and significantly shifting public perception of the conflict back home.

Colonel Tow's helicopter missions during the war earned him numerous awards and commendations, including the Distinguished

Flying Cross, Air Medal with 29 Oak Leaf Clusters, and the Republic of Vietnam Cross of Gallantry. His leadership under fire exemplified the courage and resilience required of aviators in the Vietnam War.

Despite his remarkable service, Colonel Tow remained humble, attributing much of his success to the influence of others. He fondly remembered two mentors—his father, from whom he learned honesty, integrity, and thoroughness, and General Parmer Edwards, a two-star general who taught him the importance of clear communication and decisive action.

Reflecting on his military career, Colonel Tow said, *"We were intensely concerned about the Viet Cong penetrating our perimeter. The fear and tension were always present, but we did what we had to do."* His commitment to his mission and his men never wavered, even in the face of overwhelming odds.

In addition to his military service, Colonel Tow worked for Lockheed Martin after retiring from the Army and has been an active member of the Rotary Club for 35 years, achieving perfect attendance. His impressive list of awards includes the Purple Heart, the Legion of Merit with one Oak Leaf Cluster, and the Master Aviator Badge, further underscoring his dedication and heroism.

Colonel James L. Tow's legacy is one of leadership, bravery, and service to his country. His experiences in Germany, Korea, and Vietnam reflect the complexities and sacrifices of a military career that spanned pivotal moments in modern history.

Colonel James L. Tow, thank you for your service to our great country.

Welcome Home.

Colonel Bob Walsh

United States Air Force (1968-1998)

Colonel Bob Walsh dedicated 30 years of his life to serving in the United States Air Force, including a pivotal role as a pilot during the Vietnam War. His career spanned the Tet Offensive of 1968-69, flying dangerous reconnaissance and support missions in one of the smallest planes in the Air Force's arsenal—the O-2 Skymaster. Known for his humility, class, and deep respect for his comrades, Walsh's experiences reflect the courage and determination of those who served in Vietnam.

Born and raised in Boston, Massachusetts, Walsh graduated from Boston Latin High School before attending the prestigious Air Force Academy, where he also played football. At just 25 years old, he landed at Tan Son Nhut Air Base in Vietnam. His introduction to the war was unforgettable. *"They [the Vietnamese] blew the terminal away when I got there. That was my welcome to Vietnam,"* he recalled.

From Tan Son Nhut, Walsh was sent to Da Nang, where he flew O-2 Skymasters, light propeller-driven planes used for forward air control and reconnaissance. These missions were critical for identifying enemy positions and directing ground troops and airstrikes. Flying just 500 feet above the ground—barely above the treetops—Walsh and his fellow pilots operated in constant danger. *"A great deal depended on my scouting report,"* he said. *"You were put in a position of support, and it was scary because you didn't want to fail. Our ground troops depended on us."*

Walsh spoke with deep admiration for the Marines and Army soldiers he supported. *"There is absolutely no way that I can ever express, properly, my gratitude for the Marines and Army. What they did in the A Shau Valley and in Khe Sanh was unbelievable."*

The inherent risks of flying such low-altitude missions in Vietnam were evident in Walsh's recollections. He described a harrowing incident when his plane was shot up near Da Nang. *"I lost my front prop and rear engine. I was number one coming into land. There was an F-14 behind me that lost his hydraulics. I landed in the overrun and literally saw the guy go right by me—I could have touched him. All I could think of on the approach was that I didn't want to screw up,"* he remembered.

One of the most striking memories from Walsh's service involved transporting a 4-star general to a "hot" combat zone. *"I'll never forget the general's stature—the way he got out of my plane, walking upright with distinction. Bullets were flying, and bombs were exploding, but that didn't faze the general one bit. He kept walking until he reached the bunker."*

Despite the constant danger, humor and camaraderie provided moments of relief. Walsh chuckled as he shared a story about a French reporter who was interviewing him about the war. *"She was very serious and not in favor of the war. I was supposed to fly her to Khe Sanh. All of a sudden, one of our guys comes in on a red bouncing ball with handles—bouncing into the room and onto tables. The reporter just burst out laughing."*

Although Walsh missed the iconic Bob Hope shows, he appreciated the role of USO entertainment in lifting morale. "I saw some pretty good USO shows," he said. These performances offered a brief escape from the intense pressure of combat.

Reflecting on his career, Walsh described his time in the Air Force as both challenging and rewarding. *"It was a great opportunity to serve. My first flight ever was to the Air Force Academy as a young boy. It was challenging, but like everything in life, you prepare to overcome those challenges. I was afraid to 'wash out' because I didn't want to let my parents down."*

Colonel Bob Walsh's bravery and dedication as a pilot during the Vietnam War exemplify the selflessness and skill required to serve in one of the conflict's most perilous roles. His service to his country remains a testament to the courage and determination of the Air Force's finest.

Colonel Bob Walsh, thank you for your service to our great country.

Welcome Home.

Seaman Dave Westcott

United States Navy (1964-1966)

Seaman Dave Westcott proudly served his country in the United States Navy from 1964 to 1966, spending his entire naval career aboard the USS Ingraham BD-694, a World War II-era destroyer escort that saw service through the Cold War. Growing up in Roslindale, Massachusetts, Westcott was drawn to the Navy by his love of the sea. *"I always liked boats and water, so the Navy was an easy choice,"* he said. His two years of service took him across the globe twice, experiencing the unforgiving power of the North Atlantic, the beauty of Mediterranean ports, and the camaraderie of a close-knit crew.

After completing basic training at Great Lakes Naval Base during a frigid February, Westcott received his first and only assignment on the USS Ingraham, a ship known for its reliability and versatility. *"It was a good, clean ship with a good crew,"* he recalled. Despite the grueling schedule, which left him in port for just about a week in total over two years, Westcott spoke fondly of the experience.

One harrowing memory from his time at sea involved a nighttime refueling operation in the rough waters of the North Atlantic. *"It was nighttime—you couldn't see a thing,"* he recalled. *"We had red lights to guide us. We were refueling alongside a tanker when we had a steering breakdown and had to pull away."* His ship took a dangerous 65-degree roll, a near-catastrophic event in rough seas. *"It was like a scary movie. The thing that saved us was that we had just refueled, and that kept the ship stable."*

The North Atlantic presented relentless challenges, with 30- to 35-foot seas regularly battering the ship. *"The sea was very heavy—it actually bent the frame of the bow of the ship. That shows you the power of the sea,"* he said. These conditions demanded

208

constant vigilance from the crew, particularly during dangerous tasks like refueling or navigating through storms.

Westcott's ship also conducted several Mediterranean cruises, a vital part of the U.S. Navy's presence in Europe during the Cold War. These cruises helped to maintain NATO's strategic position in the region and strengthen alliances with European countries. One of Westcott's fondest memories was arriving at a port in Madrid, Spain. *"We pulled into a large harbor with a bunch of small boats. Our ship was 325 feet long with large guns on it, and all the girls in the other boats waved. They seemed fascinated by it,"* he said with a laugh.

Despite the grueling nature of naval life, the camaraderie among the crew made it bearable. *"Our camaraderie was great,"* he said. However, not all commanding officers earned the crew's respect. Westcott recalled a frustrating incident when the ship was supposed to rescue Marines in a river in Vietnam. *"Our new commanding officer wouldn't go because he didn't want to damage the sonar dome on the ship. That turned a lot of people off,"* he said.

One memory that still haunts Westcott is a tragic training accident. *"We were practicing shooting, and planes were pulling the target. The fire control radar locked onto a plane instead of the target and hit the plane. We went to recover, and all we found in the plane was brain matter. That was not a good day,"* he said somberly.

Westcott took pride in the fact that, despite the dangers they faced, no one on his ship was killed during his service. *"Nobody was killed on our ship,"* he emphasized.

Like many Vietnam-era veterans, Westcott faced a difficult homecoming. *"It was terrible,"* he recalled. *"They told us to wear our civvies and not our uniforms. I never talked about it, except with my wife. "* The treatment of returning veterans during this period remains a painful chapter in American history, with many service members experiencing hostility or indifference.

Reflecting on his service, Westcott expressed gratitude for the opportunities and lessons it provided. *"It was a good thing. Young kids need some type of military service today. Service either makes you or breaks you. I was getting paid to see all these ports around the world—it was great!"* he said.

Seaman Dave Westcott, thank you for your service to our great country and welcome home.

Staff Sergeant Arthur Wiknik

U.S. Army 101st Airborne (1968-1970)

Arthur Wiknik served his country in the United States Army's 101st Airborne as a Staff Sergeant from 1968 to 1970. Born in Higganum, Connecticut, he was drafted at the age of 19. Despite his reluctance to join—having recently secured a job at Pratt & Whitney, bought a '68 Camaro, and fallen in love—he dutifully reported for service.

His military journey began with basic training at Fort Dix, New Jersey. *"I didn't want to go in,"* he recalled, *"but I went in."* After completing basic training, Wiknik was sent to Fort Polk, Louisiana, for Advanced Infantry Training (AIT). He described this phase as intense, with rigorous aptitude tests and weapons training. Wiknik excelled in both, which led to his enrollment in Non-Commissioned Officer (NCO) School at Fort Benning, Georgia. After graduating as a sergeant, he received additional training at Fort Gordon before being deployed to Vietnam in April 1969.

Wiknik arrived in Vietnam as the leader of a 12-man unit. Initially, he faced challenges in asserting his authority, as many of his men were battle-hardened veterans. ***"I was the new guy on the block, and no way they wanted to listen to me,"*** he remembered. However, one pivotal moment during the infamous Battle of Hamburger Hill changed everything.

The Battle of Hamburger Hill, officially known as the assault on Hill 937 in the A Shau Valley, remains one of the most grueling and controversial engagements of the Vietnam War. The 10-day battle, from May 10 to May 20, 1969, involved U.S. and South

211

Vietnamese forces clashing with North Vietnamese troops entrenched on the heavily fortified hill.

Wiknik's account of Hamburger Hill captured the chaos and terror of the battle. *"We had almost 700 soldiers ready to assault the mountain. We couldn't bring food and were told to take no prisoners,"* he recounted. Despite the harsh conditions, he managed to sneak in a can of peaches.

"I was sitting down and ate almost the whole can. A couple of guys asked me for some, but I had little left," he said. Shortly afterward, the battle erupted. *"I fired off a full magazine, three times. A shot hit the ground in front of me, covering my face with dirt, and a tracer round landed on the equipment on my chest, catching it on fire."*

In a moment of sheer courage—or madness—Wiknik yelled for his men to follow him and charged up the hill alone. *"I worked my way up to the top, turned around, and nobody was there. Nobody had followed me. I was scared to death."* His unit eventually caught up, and when he asked why they hadn't followed him initially, they joked, *"Because you didn't share your peaches."*

This act of bravery and determination earned him the respect of his men. *"After that, I had no problem with leadership,"* he said.

Holidays during his time in Vietnam held little meaning. *"The holidays meant absolutely nothing. We spent most of the time in the jungle. Christmas was just another day,"* Wiknik shared. However, he fondly remembered the camaraderie among the soldiers. *"The guys I was with were definitely a positive aspect. Everyone always looked out for each other, no matter if you were black, white, purple, or polka dot."*

Discussing the aftermath of Hamburger Hill, Wiknik became emotional. *"I just stood there, looking down, just like the movie.*

212

It hit me that I could have been killed," he said, pausing between sentences to compose himself.

Returning to the U.S. was a bittersweet experience. Wiknik hoped that the anti-war protests would help end the conflict, allowing him to return home. However, he encountered a lack of support upon arrival. *"I was ignored at the airport. People wouldn't sit next to me,"* he recalled.

Despite the indifference he faced, Wiknik remained proud of his service. *"I'm proud of the fact that I didn't do anything to embarrass my unit or family."*

He shared a poignant memory from his departure to Vietnam. *"My mom pulled me aside at the airport and showed me her wallet, loaded with $50s and $100 bills. She said, 'I'll send you to Canada with this and send you money every month to get by until the war is over.'"* Wiknik declined the offer, choosing to fulfill his duty.

Upon his return from Vietnam, the pilot of his flight announced over the intercom, *"This is my seventh return trip from Vietnam, fellas... Welcome home."*

Wiknik went on to write *Nam Sense: Surviving Vietnam with the 101st Airborne*, sharing his firsthand experiences of the war. His story has been featured on the History Channel. Married to his wife Betty for over 40 years, they have three daughters and four grandchildren.

Staff Sergeant Wiknik, thank you for your service and welcome home.

Cold War
(1947-1991)

Veterans Spotlights

Commander Carissa April – United States Navy (1993-2016), career in naval aviation and intelligence
Master Sergeant David Levesque - United States Army/Army National Guard (1979–2007)
Petty Officer 2nd Class Mike Lewis – United States Navy (1977–1994)
Rear Admiral John L. Linnon – United States Coast Guard (1957-1997)
Captain Thurman "Tom" Maine - United States Coast Guard (1983-2014)
Lieutenant Joe Ponti- United States Coast Guard (1961 – 1965)

The Cold War

The Cold War was a conflict unlike any other, the battlefield was not a distant jungle or a blood-soaked beachhead, it was the razor's edge of nuclear tension, fought in secret bunkers, aboard silent submarines, and within intelligence corridors where a single miscalculation could have meant global destruction. While they may not have stormed beaches under machine-gun fire, their sacrifice was just as profound.

The doctrine of *Mutually Assured Destruction (MAD)* dictated that if one side launched, the other would respond with equal or greater force, ensuring total devastation. It was a strategy built on fear, a high-stakes standoff where a simple mistake could mean millions of lives lost.

The Cold War is best remembered through the lens of history books and intelligence briefings. But for those who served, its realities were perhaps best captured in *The Hunt for Red October* **(1990),** a film that depicted the tension, paranoia, and brinkmanship that defined the era.

Based on **Tom Clancy's** bestselling novel, *The Hunt for Red October* told the story of a rogue Soviet submarine commander defecting to the United States, pursued by both American and Soviet forces. The film masterfully captured the secrecy and uncertainty of Cold War naval operations—where one misunderstood maneuver could lead to global conflict.

Despite the tension and secrecy, Cold War service was not just about deterrence but also about preventing instability through global presence and humanitarian efforts.

During his time on the USS Pensacola, **Petty Officer 2nd Class Mike Lewis** participated in several key missions, including anchoring off Beirut in October 1980 as part of a Multinational Peacekeeping Force.

When the U.S. Embassy was bombed, he and his fellow sailors went ashore to assist in refurbishing an orphanage outside the city. This humanitarian effort showcased the Navy's capacity for service beyond warfare. The ship's defensive measures during this tense period were equally notable. *"We ran the engines in reverse to guard against terrorists placing bombs under the ship, and Navy SEALs dropped depth charges around our perimeter,"* he explained, underscoring the constant vigilance required in such volatile environments.

For the men and women who served during the cold war, their fight was one of patience and precision, of vigilance in the face of an enemy they often never saw. They did not ask for recognition, nor did they expect it. But without their service, the world may have been a very different place.

These are their stories, and we honor them for protecting our freedoms, and fighting for the liberty of those that faced aggression and oppression around the world.

Commander Carissa April

United States Coast Guard (1993–2016)

It was a profound honor to interview Commander Carissa April, a woman whose humility and distinguished service exemplify dedication at the highest level. Serving her country for 23 years in the United States Coast Guard, she retired as a Commander, leaving behind a legacy of leadership and excellence.

A native of Freeport, Maine, Commander April graduated from Freeport High School and later attended the College of Wooster in Ohio, where she excelled as an All-American Field Hockey player in 1991 and earned Academic All-American honors in both 1990 and 1991. Her passion for the water led her to the Coast Guard, where she set her sights on becoming a professional mariner. After visiting the Coast Guard Recruiting Office, she enrolled in Officer Candidate School (OCS) in Yorktown, Virginia. Reflecting on her experience, she remarked, *"It was eye-opening. Calling it enjoyable would be a stretch, but the camaraderie and caliber of people at OCS left a lasting impact."*

Commander April's first assignment took her to Kodiak, Alaska, where she served as Operations Officer aboard a 180-foot buoy tender. It was there that she encountered one of her greatest inspirations, the late Admiral John Linnon. *"He gave me a pep talk about the importance of leadership and the weight of my position. It was a moment that stayed with me,"* she recalled.

Her journey continued with an assignment as Executive Officer on the Monomoy in Woods Hole, Massachusetts, where she oversaw

218

a 110-foot ship with expanded responsibilities in fisheries and law enforcement. Rising quickly through the ranks, she held several leadership roles, including Commanding Officer at the Regional Fisheries Training Center in Boston, Assistant Chief of the Command Center in New York (where she was on duty during Captain "Sully" Sullenberger's famous Hudson River landing), and various key positions at Coast Guard Headquarters in Washington, D.C. She concluded her career as Chief of Coast Guard Auxiliary, leading 2,500 uniformed volunteers.

Reflecting on her career, Commander April shared vivid memories of missions and rescues, often deflecting praise to her crew. One rescue, 110 miles off Cape Cod, stood out. *"Conditions were the worst I've ever seen, and everyone was seasick. We secured a wooden vessel and saved five people on board, towing it back at just 5 MPH. Camera crews were waiting at Woods Hole, but they only wanted to talk about the three women crew members and the female Ex-O. I was proud of what the team had achieved."*

She also cherished quieter moments, like a conversation with a young helmsman in the Caribbean. *"He turned to me and said, 'Ex-O, all my friends back home are doing nothing, and here I am steering this 100-foot ship.' It was incredible to witness the pride and responsibility these young people carried."*

When asked about mentors, she highlighted individuals like John Healey, who demonstrated the courage to have tough conversations and taught her invaluable lessons about communication and leadership.

Summing up her service, Commander April said, *"It was a privilege and an honor. The caliber of people, their skills, and their willingness to be part of a team—it's something I'll always cherish."*

219

Now retired, Commander April treasures her role as a mother to her two children, Max and Reese, who attend Falmouth High School. Her service to the nation is matched only by her pride in them.

Thank you, Commander Carissa April, for your remarkable service to our country.

Petty Officer 2nd Class Mike Lewis

United States Navy (1977–1994)

Petty Officer 2nd Class Mike Lewis served with pride and dedication during his 17-year career in the United States Navy. Born in North Adams, Massachusetts, he began his journey at the Navy's Recruit Training Command in Great Lakes, Illinois. Known as the Navy's only boot camp, the Great Lakes facility prepares sailors for the rigors of military life. Situated along Lake Michigan, the base experiences harsh winters, where recruits face freezing winds and icy conditions. Training at Great Lakes emphasizes discipline, teamwork, and endurance, instilling the values necessary for naval service. For Lewis, this foundational experience set the stage for an exceptional career.

After graduating from basic training, Lewis was assigned to submarine school at NAVSUBCOM Groton in Groton, Connecticut. Often called the "Submarine Capital of the World," NAVSUBCOM Groton serves as a key training hub for the Navy's submarine force. The base is steeped in history and innovation, fostering the expertise required to operate beneath the waves. Although Lewis's career ultimately led him aboard surface ships, his time at Groton provided valuable technical skills and a deeper understanding of naval operations.

Lewis's first major assignment was aboard the USS Pensacola (LSD-38), a dock landing ship based out of Little Creek, Virginia.

221

The Pensacola played a critical role in transporting Marines, landing craft, and heavy equipment, embodying the Navy's amphibious warfare capabilities. As the ship's boiler technician, Lewis faced extreme conditions in the ship's engine room. *"It was very hot down there,"* he recalled. The engine room, located deep in the ship's hull, exposed him to sweltering temperatures and deafening noise, requiring immense physical and mental endurance. *"For punishment, they sent people to the engine room—so you can get an idea of how tough it was,"* he added.

During his time on the Pensacola, Lewis participated in several key missions, including anchoring off Beirut in October 1980 as part of a Multinational Peacekeeping Force. When the U.S. Embassy was bombed, he and his fellow sailors went ashore to assist in refurbishing an orphanage outside the city. The ship's defensive measures during this tense period were equally notable. *"We ran the engines in reverse to guard against terrorists placing bombs under the ship, and Navy SEALs dropped depth charges around our perimeter,"* he explained, underscoring the constant vigilance required in such volatile environments.

Lewis's career took him on deployments throughout the Caribbean and North Atlantic, and he served on multiple ships, including the USS Dahlgren (DDG-43) and the USS Nitro (AE-23). His duties aboard these vessels further expanded his expertise in naval operations. Despite the challenges of long deployments, Lewis found camaraderie among his crewmates. Holidays at sea, though distant from family, were made special with good food, live music, and occasional performances on the flight deck. *"We missed our families, but they made it enjoyable,"* he said.

One particularly harrowing memory occurred at Roosevelt Roads in San Juan, Puerto Rico, where a tragic attack on four sailors left the crew shaken. *"A car pulled up and gunned them down. One sailor, despite being shot four times, ran back to the ship. One died, and three were critically injured,"* he recalled. The ship's whistle blew, and the crew swiftly returned aboard to get underway, demonstrating their readiness to respond to crises.

Lewis retired with honors, holding his retirement ceremony aboard the USS Massachusetts in Fall River, Massachusetts. Reflecting on his service, he expressed immense pride. ***"It taught me structure and responsibility. I'd do it all over again,"*** he said. Today, as vice president of American Legion Riders Post 125, Lewis continues to serve his fellow veterans. From providing funeral escorts to assisting with the Mobile Vietnam Wall, his dedication endures.

Petty Officer 2nd Class Mike Lewis, thank you for your service and commitment to our great country.

Rear Admiral John L. Linnon

United States Coast Guard (1957-1997)

Rear Admiral John L. Linnon Jr. retired from the United States Coast Guard in July of 1997 after a storied 40-year career spanning the globe. A native of Hartford, CT, he was commissioned through the Coast Guard Officer Candidate School (OCS) Program after several years of enlisted service. The path to becoming a Coast Guard officer is rigorous and demands a high level of dedication, academic ability, leadership potential, and physical fitness. Candidates may enter through various commissioning programs, including the U.S. Coast Guard Academy, Officer Candidate School (OCS), Direct Commission Officer Programs, or through the Coast Guard Reserve Program.

For Admiral Linnon, the journey began through Officer Candidate School (OCS), which is an intensive 17-week program designed to transform enlisted personnel and civilians with college degrees into commissioned officers. The training, conducted at the Coast Guard Training Center in New London, CT, includes leadership development, seamanship, military science, and physical fitness training. Candidates must excel in classroom instruction and hands-on field exercises that prepare them to lead in operational and command roles. Graduates are commissioned as Ensigns (O-1) and are assigned to various operational units, including cutters, aviation units, or shore-based commands.

Following his commission, Admiral Linnon served aboard seven ships, gaining extensive experience in surface operations. One of the most notable highlights of his career was his command of the high-endurance cutter Munro, based in Honolulu, HI. Under his leadership, *Munro* participated in the two-month underwater search for the wreckage of Korean Airlines Flight 007, which was shot down by the Soviet Union over the Sea of Japan. Years later, he played a critical role in the Coast Guard's response to the 1996 crash of TWA Flight 800 off the south shore of Long Island, NY.

Calling his 40 years of service *"excellent and satisfying,"* his first assignment as a flag officer was as Commander of Joint Task Force Five in Alameda, CA, the U.S. Pacific Command's center for Department of Defense support to civilian counterdrug agencies. This high-level assignment required extensive travel to Japan, Hong Kong, Singapore, Thailand, the Philippines, Mexico, and Alaska, where he strengthened international partnerships in counterdrug efforts.

Admiral Linnon's assignments ashore included serving as Chief of the Coast Guard Officer Candidate School Branch at the Reserve Training Center in Yorktown, VA, where he played a pivotal role in shaping the next generation of Coast Guard officers. He also served as the first Coast Guard Liaison Officer to the Joint Chiefs of Staff, a prestigious assignment that placed him at the heart of national security decision-making in Washington, D.C.

Other key assignments included his role as Assistant Chief of Staff for Operations for Commander, Atlantic Area, where he oversaw critical Coast Guard missions along the eastern seaboard, and his tenure as Chief of Staff of the Seventh Coast Guard District in Miami, FL, managing operations in the southeastern United States and the Caribbean.

In 1994, Admiral Linnon assumed command of the Coast Guard's First District in Boston, MA, where he was responsible for all Coast Guard operational missions in New England, New York State, and approximately one-fifth of the New Jersey shoreline.

Under his leadership, the First District responded to major maritime emergencies, protected vital shipping lanes, and conducted critical search and rescue operations in some of the nation's most challenging waters.

Admiral Linnon credited his leadership success to personally knowing his people and always making himself available. He spoke highly of his mentor, Vice Admiral D.C. Thompson, and when asked about being away from family during the holidays, he matter-of-factly stated that it was *"just part of the routine."* When the time came to retire after 40 years of service, Admiral Linnon said without hesitation, *"It was time to go."*

Over the course of his career, Admiral Linnon had the honor of meeting two U.S. Presidents, George H.W. Bush and Bill Clinton. He particularly admired President Bush and described First Lady Barbara Bush as *"a great woman."*

Admiral Linnon was a surface operations specialist, and his awards are nothing short of impressive. His decorations include the Distinguished Service Medal, two Defense Superior Service Medals, two Legions of Merit, three Meritorious Service Medals, and five Coast Guard Commendation Medals—a testament to his dedication, leadership, and impact on the service.

Rear Admiral John L. Linnon Jr. thank you for your service to our great country.

Captain Thurman "Tom" Maine

United States Coast Guard (1983-2014)

Thurman "Tom" Maine enjoyed a successful and distinguished military career in the United States Coast Guard, serving from 1983 to 2014 and retiring at the rank of Captain. Born in North Stonington, CT, he entered the Coast Guard at age 21 and was sent to Boot Camp in Cape May, NJ. His first assignment was at Coast Guard Station Provincetown, where he began a career dedicated to saving lives, protecting our shores, and serving the nation in times of crisis.

From 1994 to 1997, Captain Maine was a helicopter pilot for Coast Guard Air Station Otis at Otis Air National Guard Base. His leadership skills and piloting expertise led to his promotion as Operations Officer and Chief Pilot at the base from 2004 to 2007. Over his 31-year career, he piloted some of the Coast Guard's most elite aircraft, including: The Sikorsky H-3 (the Coast Guard's last amphibious aircraft, capable of landing on water), The Sikorsky H-60 (the most widely flown military helicopter in the world, used for high-risk rescues), and The H-65 Dolphin (a smaller, faster helicopter used for shipboard deployment and law enforcement operations).

One of the defining aspects of Captain Maine's career was his leadership during some of the most catastrophic hurricanes in U.S. history. The United States Coast Guard plays a critical role during these disasters, conducting search and rescue missions, providing emergency medical evacuations, restoring navigation routes, and delivering humanitarian aid.

Captain Maine flew rescue missions in the aftermath of Hurricane Katrina, which devastated Louisiana, Mississippi, and Alabama, with Category 5 winds and a storm surge that submerged significant areas of New Orleans.

Captain Maine also participated in missions during Hurricane Andrew, one of the most powerful hurricanes to ever hit the U.S. With wind speeds exceeding 165 mph, Andrew flattened South Florida, leaving over 175,000 people homeless.

Flying into the heart of Hurricane Andrew, Captain Maine and his team assessed damage in real-time, provided critical aerial reconnaissance, and airlifted supplies and personnel to devastated areas.

Beyond hurricane relief, Captain Maine also played a vital role in the Coast Guard's war on drug trafficking, which is one of its most underappreciated yet crucial missions. The U.S. Coast Guard is the first line of defense against drug smugglers trying to bring narcotics into the country via the ocean.

Captain Maine's career continued its upward trajectory, and in 2009, he was promoted to serve as the Coast Guard's Liaison to the Chairman of the Joint Chiefs of Staff at the Pentagon, a position he held until 2011. Following that assignment, he was promoted to Captain and selected as the Commanding Officer of the Coast Guard's prestigious Aviation Training Center in Mobile, Alabama.

During his tenure as Commanding Officer, Captain Maine faced one of the most difficult moments of his career, the loss of four Guardsmen in a tragic night training exercise.

His response to the tragedy was his proudest achievement. *"Leadership isn't just about being in command—it's about taking care of your people,"* he said. He focused on supporting the families, being constantly available, and organizing family nights to ensure they felt connected and supported.

He also successfully lobbied for the student pilot who died to receive his designation posthumously, ensuring that his sacrifice was honored.

In addition to his rescue and law enforcement missions, Captain Maine assisted the Secret Service in helicopter security operations for Presidents George H.W. Bush and Bill Clinton. When the presidents visited Kennebunkport, ME, and Martha's Vineyard, the Coast Guard provided airborne security and surveillance, ensuring their safety from any potential maritime threats.

When asked about the greatest influence on his career, Captain Maine credited the Chief Petty Officers he worked alongside. *"They taught me the importance of hands-on leadership,"* he said.

Regarding his time in the United States Coast Guard, he stated: *"The beautiful thing is that you're always doing the job it trains you to do."*

Captain Tom Maine, thank you for your service to our great country.

Lieutenant Joe Ponti

United States Coast Guard (1961 – 1965)

Lieutenant Joe Ponti served with honor in the United States Coast Guard from 1961 to 1965. Born and raised in Lawrence, Massachusetts, Lt. Ponti graduated from the prestigious United States Coast Guard Academy, where he was commissioned as an Ensign upon graduation. His first assignment took him aboard the USCG Casco, a weather ship responsible for critical patrols and search and rescue missions.

"We went on five-week patrols," he recalled. *"We carried meteorologists onboard and conducted search and rescue missions. It was demanding work but rewarding."*

During his time on the Casco, Lt. Ponti served in various roles, including 18 months as a Deck Watch Officer, six months as a Student Engineer, and a stint as an Engineering Watch Officer. His duties extended beyond the ship as well—he worked as a Marine Inspector, ensuring the safety and compliance of American passenger vessels.

Reflecting on spending holidays at sea, Lt. Ponti shared how the crew maintained their morale. *"We made the best of it. They always prepared something special, and we had very good cooks onboard. It wasn't easy, though. There were no cell phones or high-tech communication like today, so we lost contact with family for long periods."* Despite these challenges, they found creative ways to stay connected. *"One thing we did was communicate with aircraft at sea. We'd send weather balloons up*

30,000 feet to report conditions to them. Sometimes, we'd dictate postcards to the stewardesses to send home."

One memorable experience occurred in 1963 when the Casco departed from Boston on an oceanographic mission. "We collected water samples and conducted research, eventually making our way to Rio de Janeiro. That was a really nice experience," he said with a smile.

When asked about mentors during his service, Lt. Ponti humbly credited his classmates. *"Most of my mentors were the guys I served with. We had a strong bond, and our class motto was '61 Never Outdone,'"* he said, laughing at the memory.

Lt. Ponti's most significant and somber mission came during the Cuban Missile Crisis. *"In 1963, while in port in Boston, we received an emergency call. An armed guard was stationed in front of our ship. We were assigned to head out into the Atlantic as a rescue ship during the crisis. We stayed in communication with the mainland, awaiting further instructions. Eventually, we received the call to return to port and resume our normal duties."*

Reflecting on his time in the Coast Guard and his identity as a veteran, Lt. Ponti emphasized the lifelong bonds he formed. *"I've kept in close contact with my classmates for over 60 years. We call ourselves the ROMEOs—Retired, Old, Men, Eating Out. The camaraderie we share is what makes being a veteran so special. We love to joke around, compare health notes, and plan our next gatherings. These guys are like brothers to me."*

Today, Lt. Ponti channels his passion for the Coast Guard into writing and public speaking. He writes articles on Coast Guard ships for various publications and presents informative lectures at assisted living centers, sharing his experiences and knowledge with others.

Lieutenant Joe Ponti, thank you for your service to our great country and for your continued dedication to preserving the legacy of the United States Coast Guard.

Persian Gulf War (1991)

Iraq War (2003-2011)

War in Afghanistan (2001-2021)

Veterans Spotlights

Sergeant Chris Alves- United States Marine Corps (2012-2016)
Brigadier General Don Bolduc – United States Army (1981-2017)
Brigadier General John Hammond-United States Army (1983-2014)
Corporal Mitch Keil – United States Marine Corps (2006–2010)
Master Sergeant David Levesque- United States Army & Army National Guard (1979–2000)
Dr. Ron Nasif – United States Navy Reserve (1981-2011)
Brigadier General Anthony Schiavi-United States Air Force and Massachusetts Air National Guard (1983 – 2013)
Lt. Colonel Paula Smith –United States Army (1998-2018)
Lieutenant Dominick Sondrini- United States Marine Corps (2004–2008)
Command Sergeant Major Steve Valley- United States Army (1985–2015)
Master Sergeant Shawn Welsh- United States Army (2003–2023)
Sergeant Christine Zecker-United States Army (1989-1998)

Sectarian Divides and Terrorism-
An unstable Middle East

The conflicts in the Middle East, from the Gulf War to the wars in Iraq and Afghanistan have profoundly shaped the experiences of American service members. While the Gulf War demonstrated the power of advanced military technology and rapid, decisive action, the prolonged wars that followed exposed the limits of military force in achieving long-term stability.

The Gulf War

The Gulf War showcased the power of cruise missiles, stealth bombers, and satellite-guided bombs, allowing the U.S.-led coalition to dismantle Iraq's military infrastructure before ground troops even engaged the enemy.

Yet, for all its technological advancements, the war was not without its devastating effects. As Saddam Hussein's forces retreated, they set fire to Kuwait's oil fields, turning the landscape into an apocalyptic inferno. **Dr. Ron Nasif, a U.S. Navy Medical Corps** officer, recalled standing under the night sky in Kuwait, watching the flames consume the horizon. *"It reminded me how wasteful and destructive war can be,"* he reflected.

Depicting the Gulf War in the film *"Jarhead"*

One of the most striking portrayals of the Gulf War came in the 2005 film *Jarhead,* written by **William Broyles Jr.** and produced by **Lucy Fisher** and **Douglas Wick**, based on the memoir by Marine sniper **Anthony Swofford**. Unlike traditional war movies filled with intense combat, *Jarhead* captured the surreal experience of waiting for a battle.

The film follows Swofford (played by Jake Gyllenhaal) as he trains for war, only to find himself in a conflict where technology and air power do most of the fighting. Instead of engaging in firefights, he spent most of his time waiting, preparing, and dealing with the mental toll of a war.

The Iraq War

The U.S. invaded Iraq in March of 2003, in what was known as Operation Iraqi Freedom. The invasion was led by the United States, with support from the United Kingdom and other coalition forces, under the justification that Saddam Hussein's regime possessed weapons of mass destruction (WMDs) and had ties to terrorism. However, no significant stockpiles of WMDs were ever found.

Saddam Hussein's government quickly fell, with Baghdad captured by U.S. forces on April 9, 2003. After months in hiding, Saddam was discovered on December 13, 2003, near his hometown of Tikrit, in an underground bunker during Operation Red Dawn.

The War in Afghanistan

The War in Afghanistan began on October 7, 2001, as part of Operation Enduring Freedom, in response to the September 11, 2001, terrorist attacks. The United States, supported by NATO allies, launched the invasion to dismantle al-Qaeda, remove the Taliban regime, and prevent Afghanistan from being used as a terrorist base. The war became America's longest conflict, lasting 20 years until the final withdrawal of U.S. troops on August 30, 2021.

More than 2,400 US service members were killed and over 20,000 were wounded in the conflict.

Depicting the war in Afghanistan in the movie, *"Lone Survivor"*

The 2013 film *Lone Survivor*, directed by **Peter Berg**, is based on the 2007 memoir by **Marcus Luttrell,** a former U.S. Navy SEAL. The book was published by Little, Brown and Company in 2007. The film depicts Operation Red Wings, a real-life 2005 mission in Afghanistan that ended in tragedy when a four-man SEAL reconnaissance team was ambushed by a much larger Taliban force.

On June 27, 2005, four U.S. Navy SEALs—**Marcus Luttrell, (played by actor Mark Wahlberg), Michael Murphy, Danny Dietz, and Matt Axelson** were sent into the Kunar Province mountains to locate and kill Taliban leader Ahmad Shah.

While on reconnaissance, the SEALs were discovered by local goat herders. They let the herders go, unsure if they would alert the Taliban. Hours later, they were surrounded by a large Taliban force and engaged in an intense firefight. **Lt. Michael Murphy**, the team leader, exposed himself to enemy fire to call for backup but was killed in action.

Dietz and Axelson also fought to their last breath, with **Luttrell** being the lone survivor of the battle.

Severely wounded, **Luttrell** was rescued by local Afghan villagers, who protected him from the Taliban until he was recovered by U.S. forces days later.

Many who served in the Gulf War, Iraq, and Afghanistan endured multiple deployments, faced prolonged separations from their families, and battled the physical and psychological wounds of war. Post-Traumatic Stress Disorder (PTSD), traumatic brain injuries, and moral injuries became prevalent, affecting tens of thousands of veterans. The burden of war did not end on the battlefield, as many struggled with reintegration into civilian life, dealing with healthcare challenges, and finding support systems that fully understood the weight of their experiences.

Despite the withdrawal of most U.S. forces from Iraq and Afghanistan, the region remains volatile, and the long-term effects of these conflicts continue to shape American foreign policy.

The sacrifices of those who served will not be forgotten, and we honor those that served, these are their stories in their own words.

Sergeant Chris Alves

United States Marine Corps (2012-2016)

Sergeant Chris Alves' service in the United States Marine Corps reflects the dedication, discipline, and brotherhood that define the Corps. A 2012 graduate of Falmouth High School, Alves enlisted in the Marines with a clear goal: *"I wanted to be the best,"* he said proudly. His rapid rise through the ranks, achieving the rank of sergeant, is a testament to his leadership and commitment.

Alves completed basic training at Parris Island, South Carolina, where he honed the mental and physical toughness required of Marines. His role as an M1A1 Tank crewman was as challenging as it was rewarding. Heading a four-man crew, Alves worked every position on the tank, including driver, loader, gunner, and tank commander. The M1A1 Abrams, a state-of-the-art armored vehicle, required precision and teamwork to operate effectively in combat situations.

Deployed to Spain, Israel, Yemen, and Mosul, Alves experienced the complexities of modern warfare firsthand. *"It had its ups and downs, but it was good for me, despite the fact that we got shot at,"* he said. One harrowing moment occurred as his platoon prepared to move. Insurgents launched four missiles directly at their position. *"The missiles looked like big telephone poles,"* he remembered. Fortunately, two U.S. Navy ships intercepted them after detecting the launch on radar.

Operating in intense heat and under constant threat required mental and physical endurance. *"The duty was excruciating at times, especially with the heat inside the tank,"* Alves recalled. Training

239

and instinct were critical to survival. *"You train so much, the reaction kicks in, and it becomes automatic,"* he explained. When asked if he was ever afraid, his response was swift and confident: *"A Marine is never afraid."*

Despite the dangers, Alves cherished the camaraderie of his unit. *"We were a very tight-knit group,"* he said. Holidays away from home didn't faze him. *"You miss one, you miss them all. I had my buddies, which was great."* He credits his fellow Marines with shaping his character and skills. *"I had some really great guys: Darryl Brown, Romny Grimaldi, Matthew Riveira, Kelvin Thomas. Their influence ranged from having my back, keeping me in check, teaching me a variety of skills, to inspiration and motivation. Pedro Delosantos and Christoffer Ascenio were best buds too."*

Reflecting on his service, Alves called it *"the best four years of my life."* His awards, including the National Defense Service Medal, Sea Service Deployment Ribbon, Marine Corps Good Conduct Medal, Global War on Terrorism Service Medal, and Combat Action Ribbon, highlight his contributions to national security. Several additional medals are still pending.

Now pursuing a degree in computer science at Cape Cod Community College, Alves plans to use the GI Bill to complete his education. With a date set for the Police Academy in July, he's ready to continue serving his community in a new capacity.

Sergeant Chris Alves' journey from leading a tank crew in combat zones to preparing for a career in law enforcement is a testament to his resilience, leadership, and dedication. His service exemplifies the Marine Corps' values of honor, courage, and commitment.

Sergeant Chris Alves, thank you for your service to our great country.

Brigadier General Don Bolduc

United States Army (1981–2017)

You don't earn five Bronze Stars (one with Valor) and two Purple Hearts by being an average soldier. Such honors are awarded to individuals who exemplify unparalleled bravery and exceptional leadership. It was my absolute honor to interview Brigadier General Don Bolduc for Veterans Spotlight. General Bolduc's military career spanned 36 years, and his passion for his country, combined with his deep care for the well-being of his troops, is unmatched.

General Bolduc hails from Laconia, New Hampshire, a small city in the heart of the Granite State. Nestled between Lake Winnipesaukee and the White Mountains, Laconia is known for its rugged beauty and tight-knit community. New Hampshire, often referred to as the "Live Free or Die" state, has a long tradition of patriotism and service. Its motto, derived from Revolutionary War hero General John Stark, embodies the independent and resolute spirit of its people.

Growing up on a farm in Laconia, Bolduc's upbringing was steeped in hard work and discipline. *"I was born with a shovel in one hand and an ax in the other,"* he recalled, describing the early mornings spent tending to farm chores before and after school. His strong foundation, coupled with a Catholic education under the firm guidance of nuns, prepared him well for the rigors of military life. The Bolduc family had a proud tradition of service, and his grandfather insisted all Bolduc men answer the call of duty. This sense of responsibility and patriotism profoundly influenced Don, shaping him into the extraordinary leader he would become.

Bolduc's leadership potential was evident from the start. While stationed at Fort Carson, Colorado, with the 4th Infantry Division, he quickly distinguished himself, earning "Soldier of the Month" and "Soldier of the Year" honors. These achievements brought him to the attention of Brigadier General Colin Powell, who personally pinned his medal and facilitated his transfer to the prestigious 82nd Airborne Division. Bolduc described Powell as "respectful," noting how the General's approachable demeanor put everyone at ease, regardless of rank.

Bolduc's service included 81 months in combat and ten deployments, beginning with Grenada and spanning the Persian Gulf, Kuwait, and Afghanistan. His roles as a Special Forces Team Leader and commander at various levels were marked by bravery and resilience. Yet, his most challenging moments weren't on the battlefield. *"Writing letters to the families of fallen servicemen or making that call—knowing I gave the command that led to their loss—that's the hardest thing,"* he admitted somberly.

When asked about the essence of leadership, General Bolduc highlighted strong character, physical and moral courage, and compassion. *"It's not just about doing the right thing when nobody's watching—it's doing the right thing when everybody's watching,"* he emphasized. Bolduc's empathy extended to ensuring his soldiers' well-being. He prioritized sending troops home for significant family events, understanding the long-term value of these gestures. For families of fallen soldiers, he insisted someone from the unit accompany the flag-draped casket, so grieving families saw a familiar, comforting face amidst their loss.

General Bolduc's openness about his struggles with Post-Traumatic Stress Disorder (PTSD) reflects another dimension of his leadership. Combat left him with multiple brain injuries and the invisible scars of PTSD, which his wife was the first to recognize. Seeking help at Landstuhl Regional Medical Center in Germany, he pursued treatment discreetly, fearing stigma. But once he recovered, his approach shifted dramatically. He championed programs to address PTSD openly, focusing on saving lives, marriages, and relationships. *"It was ridiculous when they told me to slow down—we were making a real difference,"* he said with conviction.

New Hampshire's contributions to military service have been profound, with countless soldiers, sailors, and airmen hailing from the Granite State. Brigadier General Bolduc's story continues this tradition of excellence. Like many from New Hampshire, he embodies the values of grit, determination, and unwavering patriotism. Reflecting on his service, Bolduc remains humble. *"It was an honor to serve alongside great men and women. I'm forever grateful for the opportunity to give back to my country."*

Brigadier General Don Bolduc, thank you for your extraordinary leadership and unwavering dedication to our nation.

Brigadier General John Hammond

United States Army (1983-2014)

My first reaction when I met Brigadier General John Hammond for the first time was that he carried himself with a powerful presence, yet was welcoming, kind, and immensely helpful. When I asked if he would be willing to do an interview for my weekly column, he readily agreed.

When we met for a second time, General Hammond was once again gracious, and his passion for his country and his 30 years of distinguished military service was unmistakable. Growing up in Reading, Massachusetts, he attended the University of Massachusetts, earned his master's degree at Boston University, and later completed a National Security Fellowship at Harvard University.

His military career began when he enlisted in the Army National Guard at age 21. When I asked why he chose to serve, he said, *"I always had a plan because of the Vietnam experience. It didn't present the military in a good fashion."* He was sent to Fort Dix, New Jersey, for basic training and quickly qualified for Officer Candidate School. From there, he embarked on a career that would place him at the forefront of key operations in modern American military history.

General Hammond commanded at the platoon, troop, battalion, and brigade levels throughout all three campaigns in the Global War on Terror: *Operation Noble Eagle* (Homeland Security),

244

Operation Enduring Freedom (Afghanistan), and *Operation Iraqi Freedom*. Speaking of deployment, he said, ***"Once you understand why you're there, it becomes more important. You don't want to go, but you go because you're a soldier."***

As the commander of the 26th Maneuver Enhancement Brigade in Kabul, Afghanistan, General Hammond held an immense responsibility. He oversaw the security, life-support construction, and contracting for 11 U.S. bases, along with Title 10 activities, counterinsurgency operations, humanitarian assistance, and area support missions for U.S. forces within Kabul Province. His leadership in combat was tested in some of the most volatile environments, where decisions made in a split second meant the difference between life and death.

When asked about the key to effective leadership in battle, his response was firm: ***"You lead by example. You eat the same food, drink the same water as your men, so you have an appreciation for it."*** But he also acknowledged the brutal reality of war: ***"Being nice gets you killed."***

The highlight of his service? ***"Being a battalion commander and fighting in combat,"*** he said without hesitation.

Deployments were difficult, but nothing compared to being away during the holidays. ***"Those were the hardest,"*** he admitted. ***"You have to put your family in a box and take them out at the appropriate time. You have to stay focused at all times."*** The pressure of being in a war zone was constant, and there was little room for distraction. ***"We were strictly on water and MREs,"*** he recalled.

In 2011, General Hammond made history when he became the first Massachusetts officer since World War II to achieve the rank of general in a combat theater—a testament to his leadership.

When asked to sum up his military service, he said, *"The Army made a strong investment in my life. I like to think I gave them a good return."*

General Hammond's leadership didn't end when he retired from active duty. Today he serves as the Executive Director of the Home Base program. Home Base was established in 2009 through a collaboration between the Boston Red Sox Foundation and Massachusetts General Hospital. It's a national nonprofit dedicated to healing the invisible wounds of war for Veterans of all eras. Since its inception Home Base has developed innovative solutions for over 40,000 veterans and their families suffering from mental health (PTSD) and traumatic brain injury (TBI).

His list of military awards is extensive, reflecting a career marked by valor and dedication. Among his honors are the Army Distinguished Service Medal, the Legion of Merit Medal for combat service, the Bronze Star Medal, the Meritorious Service Medal, five Army Commendation Medals (one for Valor), two Valorous Unit Awards. Internationally, he was recognized with the French National Service Medal, a rare honor for American soldiers. General Hammond also received the American Red Cross Heroes Award for his work at Boston Hope during Covid, and in 2020 he was inducted into the US Army Military Police Hall of Fame

Boston Hope was a 1,000-bed field hospital for Covid patients that was built in 5 days. Boston Hope successfully provided care for Covid patients during the height of the pandemic.

Brigadier General John Hammond's career is one of resilience, sacrifice, and unwavering dedication to his country. His leadership in combat, his historic rise through the ranks, and his commitment to serving Veterans in their time of need stand as a testament to his lifelong devotion to duty.

General John Hammond, thank you for your service to our great country.

Corporal Mitch Keil

U.S. Marine Corps (2006-2010)

This Veteran Spotlight features Corporal Mitch Keil, a man who exemplifies class, integrity, and humility in his service to his country and his continued contributions to his community. Corporal Keil served in the United States Marine Corps from 2006 to 2010.

A graduate of Wahconah High School, Corporal Keil began his military journey at the famed Parris Island in South Carolina. Reflecting on the emotional day of his departure, he shared, *"After I graduated, I had two months before I went in. I had a chance to see my family and friends, which was great. My parents signed the papers for my induction. If they didn't, I was going to sign them anyway when I turned 18. It was an emotional day... I caught my plane in Boston, glad I got to say goodbye to my family and friends."*

Recalling his first day of boot camp, he described the experience vividly: *"The drill instructor screams at you on the bus. They keep you up for a couple of days with no sleep, giving you little power naps. You're nervous on your way there, but after the screaming, I was ready to go!"*

Following Parris Island, Corporal Keil was sent to Camp Lejeune for Military Occupation Specialty (MOS) Training, focusing on Landing Support Specialty. This role included responsibilities such as organizing beach landings and managing Departure and Arrival Control Groups. After completing his training, he joined Combat Logistics Battalion-24 for six months of intense Military Expeditionary Forces Training before being deployed to Kandahar, Afghanistan, in the Helmand Province.

248

During his deployment, Corporal Keil and his unit underwent specialized training to navigate the harsh terrain and unpredictable threats in Afghanistan. *"They sent us for specific training from Kandahar to Garmsir. They have 1,100 miles of paved road, and we must have gone through 1,000,"* he recounted. The constant state of vigilance weighed heavily on him, particularly during patrols through Kandahar City. *"It was pretty nerve-wracking. You hear stories of not knowing who the enemy is. I still get into funks when I think about what happened in Afghanistan."*

One memory that stands out vividly for Corporal Keil is an attack on his convoy. *"Our convoy was leaving Kandahar around 2:30 or 3 a.m. I was a gunner on top of a Humvee in the back. I saw a flash. An IED (improvised explosive device) hit the third vehicle in our convoy. It blew a hole in the road so big, we couldn't leave for 18 hours,"* he said. When asked if instincts and training kick in during such moments, he responded, *"It all kicks in. Whoever is behind in the convoy relays what happened. We were ready to retaliate. You wait for the word. You instantly look at the roster to see who's in the vehicle that got hit. I had my night vision goggles on, which are no help at all. We found out later it was a rogue guy trying to take out Canadian Special Forces, but he hit us instead."*

The threat of attack was a constant presence during his time in Kandahar. Corporal Keil described the unique danger posed by insurgents who used the harsh environment to their advantage. *"Insurgents would go up on the mountain every 2-3 days, dig a hole, put a rocket in the ice, and aim it at the base. They'd leave it there, and when the ice melted, it detonated the rocket and fired on our base. One landed so close, it knocked me off my bed one night. We were constantly sitting ducks."* Despite these challenges, Corporal Keil noted that morale remained high among his unit. However, some moments hit harder than others, especially holidays spent away from family. *"It was really hard being deployed over Easter. It's the only time of the year my whole family gets together. I really missed it. Nobody could get me out of that funk for a couple of days."*

When reflecting on his time as a Marine, Corporal Keil shared a profound sentiment: *"You will never see more Marines cry than on the last day of boot camp at graduation. You're full of so much pride. The emotion of it all… since I got out, that's what I miss the most: the camaraderie. I had 25 guys I served with. We were very close. We knew everyone's birthdays, siblings, anniversaries. Man, that's what I miss the most."*

After his honorable discharge, Corporal Keil attended the Massachusetts College of Liberal Arts (MCLA) on the GI Bill. He expressed gratitude for the support he received from the faculty and staff. One day, he decided to visit the North Adams American Legion, unsure of what to expect. To his pleasant surprise, he was warmly welcomed by Dennis St. Pierre, a former Commander. *"He welcomed me genuinely, introduced me around, and got to know my story."*

Today, Corporal Keil serves as the Senior Vice Commander at American Legion Post #125 in North Adams. He continues to be a respected leader in his community. His dedication was evident when he served as the Master of Ceremonies at Veterans Memorial Park last Memorial Day.

Corporal Mitch Keil, thank you for your service and commitment to our great country. Your story is an inspiration, reminding us of the sacrifices made by those who serve and the strength of the bonds forged in the military.

Master Sergeant David Levesque

United States Army/Army National Guard (1979–2007)

David Levesque served his country in the United States Army and Army National Guard as a Master Sergeant for 28 years, from 1979 to 2007. Raised in Enfield, CT, his decision to enlist was inspired by the service of numerous family members, but also by a more personal reason. *"I had wrecked my car and was getting rides, I was 19 years old, still getting rides from my mother. Originally, I went to a Navy recruitment office, but the recruiter said he was going out to get something to eat and would be back in an hour or two, didn't want to wait and within a week, I was in the Army,"* he recalled.

He completed his basic training at Fort Dix, NJ, an experience he remembered vividly. *"I was scared as hell…kept asking myself, 'What did I just do?' We were all scared, but then the camaraderie kicked in, and we started going to breakfast, lunches, and dinners together,"* he said.

His first assignment took him to Fort Devens, Massachusetts, where he trained as a Morse Code Interceptor. Fort Devens, a long-standing military post, was a key center for Army intelligence and communications training. It was here that he was required to master an intense skill set. *"You had to be able to type 90 letters per minute... It was extremely difficult,"* he recalled. However, his time at Fort Devens was cut short when a serious accident left him with head trauma, forcing a reassignment to Redstone Arsenal in Alabama.

At Redstone Arsenal, a critical hub for the Army's missile and rocket programs, Levesque transitioned to working on nuclear special weapons, including Trident and Pershing missile systems. *"It was very fun, but we all knew the dangers of it. The Chief Warrant Officers were great mentors,"* he said. His time at Redstone placed him at the forefront of one of the most technically advanced and strategically significant fields in the military.

When asked about holidays in the service, he responded, *"The first one was tough, but it didn't really affect me. Wrote and received a lot of letters and I was single, and took duty for the married guys, so it didn't matter... really felt the camaraderie, that was important."*

Levesque's first deployment to Iraq put him in the middle of a highly intense combat environment. *"We got mortared every other day. The mortars hit so close that you feel the vibrations on the ground, and you realize how fragile we really are,"* he said. *"I thought about my wife and kids. I had a few close calls... It gives you a reality check, big time."* He spoke about the mental toughness required in such situations. *"Your mindset needs to be in the present. It's unfortunate, but you need to forget about the family back home or it becomes extremely dangerous for you and the people that you're leading."*

As a senior leader, he made it a priority to check in on the soldiers under his command. *"In my position, I had the chance to go around and talk to other soldiers about their experiences and see how they were feeling. In combat, you can't keep it in, you need to talk about it, so I'd find my guys and we'd talk, communicate. If you keep it in, that's when you snap."* He was grateful that all

150 soldiers under his command made it home. *"Some were injured, but they came home and that means a great deal to me."*

When reflecting on his service, Master Sergeant Levesque summed it up in one word: *"Pride, no question. I am very proud of my 28 years of service, the people I've touched and mentored and the people that have mentored me. The pride is always there, which makes the camaraderie of a combat veteran so special."*

Throughout his career, he was awarded the Bronze Star *("I was just doing my job")*, the Good Conduct Medal, the Meritorious Service Medal, and numerous service medals.

Master Sergeant Dave Levesque, thank you for your service to our great country.

Commander Dr. Ron Nasif

United States Naval Reserve (1981-2011)

Commander Dr. Ron Nasif dedicated 30 years of service to the United States Naval Reserve, balancing a successful medical career with his duties as a Naval officer. A native of Boston, Massachusetts, he majored in Biology at Boston College, where he also played football for the Eagles. He went on to complete his medical education at Tufts Medical School and later specialized in orthopedic surgery.

Dr. Nasif accepted his commission in 1981, embarking on what he described as a "learn on the job" experience. *"I would have been totally lost without my Corpsmen,"* he recalled. *"They taught me a great deal—how to salute, where to bunk, how to order supplies. I'm forever grateful to them."* He trained as an Aviation Medical Officer in Pensacola, Florida, authorizing pilots for flight duty and learning the intricacies of aircraft operations.

Throughout his career, Dr. Nasif served in various naval hospitals across the country and abroad. His dedication to service extended to two major deployments—the first during the Persian Gulf War and the second during Operation Iraqi Freedom. His experience in Kuwait during Operation Iraqi Freedom left a lasting impression on him.

Stationed at a base in Kuwait, Dr. Nasif and his team worked closely with local civilians, teaching them how to use gas masks in the event of a chemical attack. *"The language barrier made things challenging,"* he said. *"But then they brought in about 30 kids, and everything changed. A supply plane had come in with bicycles. We gave them to the kids, and they were so happy—they kept saying, 'We love America.' It was a beautiful sight."*

Despite the heartwarming moments, Dr. Nasif witnessed the harsh realities of war. He recalled standing under the night sky in Kuwait, watching the flames from the oil wells Saddam Hussein had set ablaze. *"It reminded me how wasteful and destructive war can be,"* he reflected.

The emotional toll of war wasn't lost on Dr. Nasif. He spoke candidly about the lives lost during his deployments and the burden of carrying those memories. *"I realized early on that we in the military weren't making policy. We were there to do our jobs. I would tell my colleagues not to give their opinions on the war because it wasn't our place. But it still bothers me—the lives we lost over there."*

Throughout his time in the Navy, Dr. Nasif credited his commanding officers for their mentorship and support. *"They made a tremendous effort to integrate reserve officers like myself,"* he said. *"I was always called on to fill the gap when needed, and they respected that. I respected them just as much."*

Dr. Nasif's reflections on his service are filled with both pride and humility. *"It was an honor to serve the greatest country in the world,"* he said. *"It was a wonderful experience that shaped who I am today."*

Today, Dr. Ron Nasif continues his service to veterans as a member of the Veterans Council in Falmouth and the Disabled American Veterans (DAV). His commitment to supporting his fellow veterans underscores his lifelong dedication to service.

Dr. Ron Nasif, thank you for your service to our great country.

Brigadier General Anthony Schiavi

United States Air Force and Massachusetts Air National Guard (1983 – 2013)

This Veteran Spotlight of Brigadier General Anthony Schiavi took place over a decade ago in his office at the 102nd Intelligence Wing at Otis Air National Guard Base on Cape Cod. I was there to discuss plans for a Hometown Hero's event, aimed at entertaining over 300 troops. My first impression of then-Colonel Anthony Schiavi can be summed up in one word—Impressive. That impression has only deepened over the years as General Schiavi has become a trusted mentor and a steadfast supporter of my work with active military and veterans.

Otis Air National Guard Base, situated within Joint Base Cape Cod, is a hub of military history and strategic importance. The 102nd Intelligence Wing, in particular, has been integral in supporting national security efforts, making it a fitting post for a leader of General Schiavi's caliber.

General Schiavi's career spanned 30 distinguished years in the United States Air Force and the Massachusetts Air National Guard, culminating in his retirement from active service in 2013. Before retiring, he served at the Joint Force Headquarters of the Massachusetts National Guard and became the inaugural Executive Director of the Massachusetts Military Reservation.

Born in Framingham, Massachusetts, General Schiavi earned his mathematics degree from Assumption College in 1983 and graduated from Squadron Officer's School at Maxwell AFB in Alabama in 1989. His assignments were as diverse as they were prestigious, including Undergraduate Pilot Training at Laughlin AFB in Texas, F-15 Pilot duties with the 58th Tactical Fighter Squadron at Eglin AFB in Florida, and deployment with the F-15 33rd Tactical Fighter Wing Provisional at Tabuk Air Base in Saudi Arabia. He later served as an F-15 instructor pilot and Chief of Training at Otis Air National Guard Base.

Being an F-15C pilot is a calling that demands precision, discipline, and exceptional skill. Flying the F-15 Eagle, a legendary air superiority fighter renowned for its speed, maneuverability, and advanced weapons systems, requires relentless focus. The experience of piloting an F-15C isn't just about flying—it's about mastering a powerful machine built for dominance in the skies, making decisions under pressure, and carrying the responsibility of protecting national airspace.

General Schiavi's combat record speaks volumes about the demands of the role. A command pilot and veteran of Operation Desert Storm, he flew 56 combat missions and secured one confirmed aerial victory. His flightlog boasts over 2,800 hours and 308 combat hours across three tours in Southwest Asia. Reflecting on his experience, General Schiavi shared the reality of aerial combat: *"I was a relatively new F-15C pilot when we got the call to deploy to Saudi Arabia after Iraq invaded Kuwait. We spent six months during Desert Shield preparing for what would become Desert Storm, facing off against the fourth-largest air force in the world and one of the most heavily defended capitals. Our training instilled a tremendous sense of confidence, allowing us to control our adrenaline in high-stress moments. The reality only hits when the missiles start coming off the jets—that's when you realize it's not training anymore. But you manage your emotions, just as you're trained to do."*

Leadership, according to General Schiavi, hinges on authenticity and trust: *"You have to make tough decisions, but showing empathy and compassion during those moments earns the*

257

respect of your team." He acknowledged that the greatest challenge of command lies in carrying the weight of responsibility for both mission success and the well-being of the people under your command. *"Sending your people off to deployment, knowing the difficulties they and their families face, and praying for their safe return—that can cause sleepless nights. But you ensure they're ready and receive the best training possible."*

General Schiavi recalled the 9/11 attacks as one of his most defining experiences. *"Being a commander during those attacks and witnessing our unit respond with such precision and professionalism under unimaginable circumstances filled me with pride. We were the nation's first military response that day and watching everyone rise to the occasion was awe-inspiring."*

His accolades are extensive, including the Legion of Merit, Distinguished Flying Cross with Valor Device, Meritorious Service Medal with one oak leaf cluster, Air Medal with four oak leaf clusters, Kuwaiti Liberation Medal (from Kuwait and Saudi Arabia), and many more awards that reflect his dedication and exemplary service.

General Anthony Schiavi, your service to our nation stands as a testament to courage, leadership, and sacrifice. Thank you for your unwavering commitment to protecting our freedom.

Lieutenant Colonel Paula Smith

United States Army (20+ Years of Service)

Lieutenant Colonel Paula Smith dedicated more than two decades to serving her country in the United States Army, retiring with an impressive career of leadership, perseverance, and dedication to both her profession and her family. Growing up in Woods Hole and attending Falmouth Public Schools, she pursued higher education at the University of Iowa. A daring application to the US Army Baylor Graduate Program for Physical Therapy led to her selection as one of only 13 candidates, earning her a full commission as a 2nd Lieutenant.

Her military journey took her through various critical assignments, including Fort Riley, KS, where she achieved multiple promotions and started her family. She served in Korea, completed Air Assault School at Fort Campbell as the only woman in her class, and conducted basic training at Fort Jackson, SC, where she pursued her doctorate and started an Airborne Orientation Course. By the time she was assigned to Fort Stewart, GA, Lt. Colonel Smith had become a vital medical officer, leading 13 surgical services and providing essential training within the 1st Cavalry Division Brigade Combat Team.

In 2008, Lt. Colonel Smith deployed from Fort Hood, TX, to Northern Iraq as a Physical Therapist for a Brigade Combat Team, responsible for the health and recovery of over 3,000 soldiers. In a combat zone like Mosul and Tal Afar, her role was anything but conventional. Physical therapists in such environments did much more than rehabilitation—they were often the only providers of musculoskeletal care in remote outposts, helping soldiers recover from injuries sustained during missions, harsh terrain, and grueling training.

Her work required resilience, adaptability, and the ability to operate in austere, high-pressure environments. *"I lived out of an assault pack,"* she recalled, navigating the unpredictability of combat zones where medical professionals often had to move with frontline troops, setting up makeshift clinics wherever necessary. In these settings, she treated injuries ranging from sprains and fractures to more severe cases caused by combat. Physical therapy was crucial for maintaining combat readiness, as untreated injuries could lead to decreased operational capacity.

Lt. Colonel Smith's experience was also marked by her service as a Eucharistic Minister, providing spiritual support to soldiers in need. Despite the danger and uncertainty, she emphasized the strong bond within her unit. *"The 1st Cav was just like family… it was a really good family vibe,"* she reflected, her voice filled with emotion.

One of her most significant challenges was balancing her identity as both a soldier and a mother. *"I struggled with being a mom in a non-traditional setting—in a highly dangerous war zone, combat environment,"* she shared. Her commitment to her children remained unwavering, and she worked to maintain a sense of normalcy even in the most abnormal circumstances. *"I always did the family stuff with my kids… always emphatic to have dinner with my kids… wanted my kids to value my service."* She paused, her voice heavy with the weight of memory. *"My son said to me, 'Mom, if you die, I want to die with you.' My kids are strong… they went through a lot… they're still very proud of what I've done and what service members go through."*

Returning from deployment, Lt. Colonel Smith continued to serve her country by working at The Pentagon for the Soldier For Life program under the Chief of Staff of the Army. *"It was great... got to go to the White House and Congress,"* she recalled with pride. *Reflecting on her service, she spoke with deep emotion: "It was an overwhelming honor... It was such a good fit... I don't regret a single day... I learned from the Army, and the Army learned from me."*

Lt. Colonel Smith's extraordinary career is marked by numerous accolades, including the Bronze Star, the Order of Military Medical Merit Award, the Army Surgeon General's Distinguished Military Occupational Specialty Proficiency Award, and the Iron Major Award. Her legacy stands as a testament to the vital role medical professionals play in the military and the unwavering dedication of those who serve in both combat and support roles.

Lieutenant Colonel Paula Smith, thank you for your service to our great country.

Lieutenant Dominick Sondrini

United States Marine Corps (2004–2008)

Dominick Sondrini served his country for four years as a Lieutenant in the United States Marine Corps. After graduating from Western New England College, he attended Officer Candidate School in Quantico, Virginia, where he also completed the prestigious Infantry Officer Course. *"Getting through these felt like a massive accomplishment... the drill instructors, some were mean, some were funny... there's so much yelling, but I found lots of elements funny,"* he recalled. Lt. Sondrini shared a particularly humorous memory: *"This drill instructor is in my face, screaming that I stink... tells someone to get him a can of aerosol... then sprays it all around me,"* he said with a smile.

His first deployment began on June 6, 2006. *"I was pretty scared... didn't know what to expect... it was a very emotional time with my family,"* he admitted. The deployment was with the Marine Expeditionary Unit. *"We got on Navy ships... went to Italy, France, Jordan, Kuwait... we were out on the water a week or two at a time,"* he recalled. After returning safely to the U.S., Lt. Sondrini attended the Winter Mountain Leaders Course in Sierra, Nevada.

His second deployment took him to Fallujah, Iraq, where his unit worked closely with the Iraqi Highway Patrol. *"The takeaway is looking at all the cities that were destroyed... but I remember young Iraqi kids playing soccer amongst the ruins, like they were on a regular field,"* he reflected. When asked about being deployed during the holidays, he acknowledged, *"There's no*

262

getting away from it—it was lonely. I was close with my Platoon Sergeant and the guys in my unit... camaraderie is what kept us going, plain and simple."

Lt. Sondrini recounted a defining moment from his deployment: *"In 2007–2008, there wasn't a determined enemy. Every mission put us in situations where we could harm someone... we were always aware of that. This one time, a vehicle kept approaching our convoy and didn't follow our warning protocols... we shot up his car but didn't kill him. The interpreter explained that the man couldn't read. We left him there with two bottles of water and a box of donuts. The interpreter was the most important person in our platoon."*

Interpreters played an indispensable role in Iraq, not only serving as linguistic bridges between U.S. forces and Iraqi civilians but also providing critical cultural insights that often determined the success or failure of an operation. Many interpreters risked their lives daily, navigating dangerous terrain alongside American forces. They were often the first to identify potential threats, defuse tense encounters, and provide invaluable context to Marines like Lt. Sondrini, ensuring that communication breakdowns did not escalate into unnecessary violence. Without them, the challenges of operating in a foreign land with deep-seated conflicts would have been exponentially more difficult.

Lt. Sondrini shared another powerful memory: *"We ran a checkpoint... civilians had a strict 10 p.m. curfew... we shut down the highway... had 20-25 people screaming and yelling... it was getting ugly. The Iraqi Highway Patrol went down the highway and began firing their guns into the air, which only escalated the tension... people finally calmed down when they brought out the attack dogs,"* he recalled.

When asked about the best part of his service, Lt. Sondrini responded, *"You're presented with very difficult challenges, and you're able to work through them. I took my Marine experience into everyday life." As for returning home, he reflected, "I didn't feel good or bad… it was kind of a culture shock… but I had a wonderful reception. I have great family and friends."*

Lieutenant Dominick Sondrini, thank you for your service to our great country.

Command Sergeant Major Steve Valley

United States Army (1985–2015)

Command Sergeant Major Steve Valley dedicated 30 years of service to the United States Army, retiring with a wealth of experience from multiple deployments and leadership roles. Growing up in Brockton, Massachusetts, he got his first taste of the military in North Adams while attending college, serving in the A-2 104th Infantry Division. His journey took him to Fort Benning, Georgia, for basic training and Advanced Infantry Training. *"I like to be challenged... physically, I knew I could do it since I played sports in high school... but I don't think anyone can prepare you for the intense mental aspect, where you have three people yelling at you at the same time... they break you down to the lowest level, then build you up so you feel bulletproof,"* he recalled.

His first assignment as an infantryman took him to North Adams before he transferred back to Brockton and became qualified in field artillery. Over the course of his career, he deployed six times—once to Iraq, twice to Japan, and three times as a civilian working for CENTCOM. His time in Iraq left a lasting impression. *"It was the wild, wild west for sure... everything that could go wrong did. I was in the Green Zone, and we got bombed constantly by the enemy. We also traveled on the world's most dangerous highway—Route Irish. Every day you were training for something to happen... it was an active war zone. Rockets*

265

were fired at us at least five times a week coming out of the chow hall... you could hear the gunshots, sometimes bullets whizzing by your head. We called it 'Full Battle Rattle'—you could never get complacent," he said.

Despite the constant danger, some missions stood out more than others. Patrols in downtown Baghdad were among the most harrowing experiences for U.S. forces. These missions placed soldiers in direct contact with insurgents and exposed them to the unpredictability of urban warfare. *"I was glad I wasn't a door banger... was happy to not have to go to downtown Baghdad... those were the real heroes,"* he admitted. Navigating through the congested streets, dealing with hostile forces hiding among civilians, and countering IED threats made every patrol a test of endurance and skill. The risk was ever-present, but so was the camaraderie among soldiers who relied on each other for survival.

When asked about fear, Command Sergeant Major Valley surprised me with his answer. *"The only time I was afraid was in Japan... they had an enormous earthquake, like a 9+ magnitude. I was never afraid of bombs, bullets, or warfighting, but I was afraid when that earthquake hit... felt completely helpless,"* he remembered.

Spending holidays overseas was another challenge. *"We worked... missed home... I had a wife and two small kids. No one can train a spouse for everyday life... they are the glue that holds it all together. Holidays were tough but definitely tougher for the loved ones back home. As a First Sergeant, I made sure everyone got a phone call home... tell your family you're OK and that you love them at the end of the day,"* he recalled.

Mentorship played a significant role in his career. *"Had one of the world's best—Jordan St. John... Marine, Vietnam vet... took a personal interest in me. To this day, one of my closest friends. For me personally, as an officer, it was about putting in the effort with my men—communicating and getting to know them,"* he said.

Reflecting on his military experience, he shared what stood out the most: *"The teamwork... when I was an Army Reservist, earning the respect of active-duty military members was important—you had to prove you were equal and fully competent. Always a soldier first, your specific skill set was secondary."*

On his service to the country, he expressed deep pride. *"Felt it was an obligation... it takes a sense of pride when you go across the world... we have our faults, but we're still the greatest country in the world. It was an honor to serve. The military is not for everyone, but everyone should be on the docket to pay it back,"* he said.

Beyond his military service, Command Sergeant Major Valley is a proud father of two officers—Captain Alex Valley of the Army Reserve and Captain Sam Valley, a U.S. Marine Corps Osprey pilot. He also captured his experiences in his book, *Inside The Fortress: A Soldier's Life Inside The Green Zone*, published in 2009 by American Book Publishing Group, Valley offers a detailed account of his life experiences within Baghdad's Green Zone, Iraq's most fortified zone during the war.

Command Sergeant Major Steve Valley, thank you for your service to our great country.

Master Sergeant Shawn Welsh

United States Army (2003–2023)

Shawn Welsh served his country in the United States Army, retiring as a Master Sergeant after 20 years of service. Born in Puerto Rico and raised as a Navy brat, military life was always familiar to him. After graduating high school and college, he joined the Army at the age of 28. His basic training took place at Fort Jackson, South Carolina. *"I was twenty-eight and older than most... most kids in basic had never even heard of the movie* **Stripes**... *getting back in shape was a challenge... the whole process is designed to break you down and instill military discipline... really not that hard if you do what you're told,"* he recalled.

Master Sergeant Welsh's first assignment was with the 4th Brigade of the 1st Armored Division in Germany. *"Germany was awesome... beautiful country, amazing food, and phenomenal beer... the fact that you made an attempt to speak the language went a long way with the German people,"* he remembered. His first of four deployments would take him to Iraq. *"It was a high-combat zone... very eye-opening... we took a lot of rockets on base... many guys struggled with temporary PTSD... completely changed my outlook on life,"* he said.

His second deployment took him to the volatile city of Ramadi. At the time, Ramadi was one of the most dangerous places in Iraq, often considered the epicenter of the insurgency. The city, located in the Anbar province, saw relentless attacks by insurgents using roadside bombs, ambushes, and sniper fire. U.S. forces engaged in intense urban combat, as they worked to regain control of the city from al-Qaeda in Iraq. The region's instability made every mission unpredictable, and for Master Sergeant Welsh, the cost of war was personal. *"I had a child during my second deployment... a child grows up without a father... those things weighed on you constantly... lost friends... that sort of sticks with you... had a friend who I lost when a roadside bomb exploded... you never get over that,"* he shared.

Despite the hardships, there were moments of camaraderie and even celebration. When asked about spending the holidays while deployed, he said, *"Honestly, some of my favorite memories... I was deployed when President George W. Bush showed up to serve Thanksgiving... you focus on where you are and what you're doing... you have your second family, and you get through the situation together."*

Losing fellow soldiers was one of the hardest aspects of service. *"I think everyone should experience the ceremony because of the emotion and honor... then again, I hope nobody has to go through it... attended more than I ever care to... don't know if there are words to describe it... sitting there and watching the care and honor that goes into it... it's an experience like no other,"* he said.

Was he ever afraid? *"Absolutely... especially on my first deployment... the first time a rocket goes off, it gets your attention... fear played a major part in several different aspects,"* he admitted. When asked about a mentor, Master Sergeant Welsh spoke with admiration: *"Master Sergeant Maurice Cole... taught me what it meant to take care of soldiers... he was a former infantryman... made me a better leader."*

What is his definition of a leader? *"Someone who leads through actions and not words... has to be empathetic and compassionate... has to be good under pressure... as Master Sergeant Cole used to say, 'Pressure busts pipes.'"*

Reflecting on his service, he said, *"It was an honor... grew up around the military... 9/11 was this generation's Pearl Harbor... thought it was my duty to serve my country... an honor to be able to serve and continue to serve the veteran community."*

Now retired, he continues his mission of supporting veterans as the host of the *Vet SOS Podcast.*

Master Sergeant Shawn Welsh, thank you for your service to our great country.

Sergeant Christine Zecker

United States Army (1989-1998)

Sergeant Christine Zecker's journey in the United States Army began with an enlistment on Halloween night at the age of 21. Her military career was distinguished by her unique and challenging role as a psychiatric specialist, where she worked to help soldiers and veterans cope with the psychological toll of war, particularly PTSD (Post-Traumatic Stress Disorder). Her service took her from basic training at Fort Jackson, South Carolina, to the psychiatric unit at the Audie Murphy Veterans Hospital in San Antonio, Texas, where she made a lasting impact on the lives of those she served.

Reflecting on her early days in the Army, Zecker recalled the intensity of her basic training experience. *"It was the first real basic training on the verge of war. Lots of people went AWOL or had breakdowns. We were in lockdowns with newspapers and media outlets. They wouldn't let us see anything that would distract us from our training."* Despite the pressures, Zecker excelled, serving as Platoon Leader for her entire class.

After completing Psychiatric Specialist Training at Fort Sam Houston, Zecker was assigned to the psychiatric unit at Lackland Air Force Base. There, she worked with veterans, dependents, and other patients dealing with PTSD, drug and alcohol addiction, and depression. Her role required compassion, resilience, and a deep understanding of the psychological wounds inflicted by war.

"We saw a lot of crazy stuff," she said. One memory that stayed with her involved a Vietnam veteran who had been a Tunnel Rat—

271

a soldier tasked with entering and clearing enemy tunnels. *"He went outside in the courtyard, smoking a joint, and had a razor blade with him. He was talking to himself. He had that look—it's unforgettable. The war did that to him. He couldn't function in society. You can't see your best friend's head on a stick and come out alright."* Zecker managed to talk the man down and retrieve the razor blade, but the experience left a deep impression on her.

Despite the heavy emotional toll of her work, there were moments of levity. *"We had this guy who all he talked about was knowing the president and how he embezzled all this money. And you know what? The Feds actually came and got him,"* she laughed.

When asked about fear, Zecker admitted that certain aspects of military life were daunting. *"I could throw a grenade or fire my M-16 no problem, but jumping off a 60-foot tower was petrifying. Crawling under barbed wire while being fired at with live ammunition was equally terrifying."*

One of the toughest parts of her job was working with soldiers preparing to deploy. *"We helped the Huey Pilots a lot with guided visualization,"* she explained. Her unit specialized in helping soldiers manage the mental strain of war. *"Our unit was trained to prevent PTSD, but you never prevent PTSD. You try to help deal with it. Our unit was one of only six to specialize in PTSD,"* she said proudly.

The holidays were particularly challenging during her service. *"In basic training, they canceled Christmas. It was a vulnerable time with the war. I was on KP duty, cleaning pots bigger than me. I cried a lot,"* she recalled. However, a simple act of kindness from a senior officer lifted her spirits. *"A General came in and wished me a Merry Christmas, asked me where I was from. My mom was a baker and sent our platoon cookies and presents."*

272

Zecker also shared fond memories of the support she received from home. *"My dad would walk around town and take pictures of things and send them to me. Mom sent clippings from the local newspapers on people in town who were deployed. It was nice,"* she said.

Looking back on her military career, Sergeant Zecker described her service as one of the greatest honors of her life. *"It was the greatest privilege and honor, aside from being a mother. I'm very patriotic and love our country so much. I'm still close with the women I shipped out to basic with, as well as my unit."*

Sergeant Christine Zecker's work as a psychiatric specialist helped countless veterans navigate the aftermath of combat and trauma. Her role in addressing PTSD within the military community reflects the importance of mental health care for soldiers, both during and after their service.

Sergeant Christine Zecker, thank you for your service to our great country.

Veteran Spotlight Index

About the Author

Wayne Soares is an actor, entertainer and tireless Veteran advocate. The former ESPN Radio Broadcaster also entertains our military women around the globe.

Wayne will begin his 3rd season as the host and producer of the popular veteran television cooking show, The Mess Hall. In 2024, he created and produced the immensely powerful Vietnam Veterans Documentary, "Silent Dignity." He is currently co-producing episode 2 with award winning actor Armand Assante. He also pens a nationally syndicated column titled Veterans Spotlight.

Wayne will be gearing up for the 7th annual Wayne Soares Military Golf Classic in 2025, which benefits his Veteran meals program, wheelchair project and clothing initiative. He continues to devote his time to VA visits, coffee with vets and hosting Veteran social events and receptions.

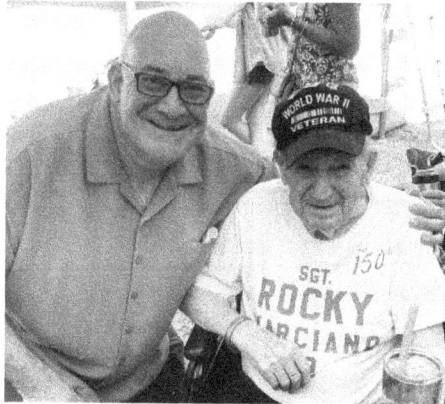

With my pal Cheeso, age 102. An American hero and patriot.

www.ingramcontent.com/pod-product-compliance
Lightning Source LLC
Chambersburg PA
CBHW021220090426
42740CB00006B/310